To: Katherine Lee

From: Prof. McCartn

Outstanding work in Math 204!

GNOMON

MIDHAT J. GAZALÉ

GNOMON

From Pharaohs to Fractals

PRINCETON UNIVERSITY PRESS

PRINCETON, NEW JERSEY

Library of Congress Cataloging-in-Publication Data

Gazalé, Midhat J., 1929–
    Gnomon : from pharaohs to fractals / Midhat J. Gazalé.
        p.   cm.
    Includes bibliographical references and index.
    ISBN 0-691-00514-1 (cloth : alk. paper)
    1. Similarity (Geometry).   2. Spirals.   3. Fractals.   I. Title.
QA447.G39   1999   516.2—dc21   98-49406   CIP

This book has been composed in Berkeley Book

The paper used in this publication meets the minimum requirements
of ANSI/NISO Z39.48-1992 (R 1997) (*Permanence of Paper*)

http://pup.princeton.edu

Printed in the United States of America

1 3 5 7 9 10 8 6 4 2

To ▲ Stéphane,

▲ Valérie,

and ▲ Olivia

Contents

CONTENTS

Preface ▲

During my student years at the Cairo, Egypt, Faculty of Engineering, I became fascinated by Bernoulli's miraculous spiral (though not to the same mystical degree as Bernoulli) and was intrigued by its kinship with the phase portraits of the damped oscillatory systems taught in college and engineering schools. Why, I asked myself, are the equations governing the oscillation of the simple pendulum exactly the same as those of the inductance-capacitance-resistance circuit, albeit with different physical meanings attached to the mathematical symbols?

Surely, energy inexorably escapes the physical system under consideration, which eventually comes to a standstill. But why do the two chosen protagonists—namely position and velocity in one case, voltage and current in the other—race one another in a manner that can be captured by a spiral phase portrait?

Like many tinkerers, I played with the Fibonacci sequence and wondered about its kinship with continued fractions and spirals. I had also come upon the design of certain classes of electrical filters using continued fractions. In my spare time, I assembled little devices, using strings and pulleys, that could calculate the square root of 2 and convert a number from binary to analog and vice versa. Having become somewhat familiar with continued fractions in connection with the calculation of surds, it turned out that the so-called whorled figures, the precursors of logarithmic spirals, were perfect metaphors for these unusual fractions.

I also fell victim to the charms of the golden section and its innumerable ramifications, and I became excited by the discovery of the Padovan number through an article in *Scientific American* by Ian Stewart.[1] In playing with equilateral triangles, I discovered a strange little pentagon that shares some interesting properties with the golden rectangle; and that I have it named the silver pentagon.

Upon discovering Benoît Mandelbrot's *The Fractal Geometry of Nature*[2] and *The Beauty of Fractals* by H. O. Peitgen and Peter H. Richter,[3]

[1] Ian Stewart, "Tales of a Neglected Number," (*Scientific American*, June 1996) pp. 92–93.

[2] Benoît B. Mandelbrot, *The Fractal Geometry of Nature* (New York: W. H. Freeman, 1981).

[3] Heinz-Otto Peitgen and P. H. Richter, *The Beauty of Fractals* (Heidelberg: Springer-Verlag, 1986).

I remembered my doctoral thesis of 1959, in which I had made extensive use of the Kronecker product, and found that the product could elegantly generate simple fractals, those self-similar patterns par excellence.

I could not help seeing logarithmic spirals everywhere! The spiral seemed to epitomize self-similarity, for which I coined the word *gnomonicity*. The central notion in this book is precisely that of *self-similarity*.

Hero of Alexandria defined the *gnomon* as that figure (a number or a geometric figure) which, when added to another figure, results in a figure similar to the original.

Chapter I is dedicated to the term *gnomon* itself, its usage and its history. I begin the book with a historical perspective of the gnomon in ancient Greece, dispel some widespread erroneous views of obelisks as gnomons, and trace the first use of the term *gnomon*, with the connotation of "that which allows one to know," to the *setchat*, or *merkhet*, of the ancient Egyptians.

Chapter II examines figurate numbers, which inspired the Greek notion of gnomon and number similarity. A discussion of *m*-adic numbers follows, and that of Hamiltonian paths upon three and four-dimensional solids, with practical applications such as binary (dyadic) and ternary (triadic) Gray codes and encoding wheels, as well as the discussion of such classical games as the Tower of Hanoi and the baguenaudier.

Central to the study of self-similarity are those sequences I refer to as Fibonacci sequences of order *m*, where $F_{m,n+2} = F_{m,n} + mF_{m,n+1}$. Explicit formulations are given for the *n*th term of these sequences, and their kinship with continued fractions is examined. Finally, the relationship between Fibonacci sequences and hyperbolic as well as trigonometric functions is established when *m* is very small.

A practical application of continued fractions and Fibonacci sequences takes the form of Ladder networks, consisting of cascades of "transducers." The reader is then introduced to electrical ladder networks. Surprising as that may seem, the analysis gradually evolves from continued fractions to the equation governing wave propagation along a transmission line. Finally, a section is dedicated to ladders made up of pulleys.

Chapter V studies in detail the so-called whorled figures, the precursors of logarithmic spirals. Monognomonic and dignomonic whorled figures are introduced, the figures being rectangles or triangles.

Chapter VI is dedicated to the famous golden section, about which so much has been written. I present an approach predicated upon whorled figures, together with a historical perspective, avoiding the pitfalls of the widespread golden number mysticism.

Following the work of Padovan, I introduce an original shape, which I call the "silver pentagon," and which possesses mathematical properties akin to those of the golden rectangle. Chapter VII closes with a strange little figure, the *winkle*, which was originally dicussed by S. Golomb under the name "Rep-Tile," and which Kronecker would have probably dismissed as nonfinitary.

No figure epitomizes self-similarity better than Bernoulli's miraculous logarithmic spiral. A step-by-step analysis of progressively more complex rotation matrices is offered, which depict the behavior of the logarithmic spiral. Finally, I present the general solution of damped oscillations, with its characteristic spiral phase portrait. The results are applied to the study of the simple pendulum and the electrical resistance-inductance-capacitance circuit, without resorting to analysis, but strictly using finite difference methods.

No book on self-similarity can close without a discussion of fractals. Much has been written about fractals since the seminal work of Mandelbrot and its ramifications into practically every realm of human endeavor, from the so-called exact sciences to the softer human sciences. It is not my intention, therefore, to write one more course on fractals. Rather, the approach presented in the last chapter is somewhat unusual, as it derives directly from number-theoretical considerations. I offer a selection of original fractal figures.

I cannot close without paying a special tribute to those individuals who unknowingly influenced me the most. First and foremost is Martin Gardner, whose acumen, wit, and unselfish kindness have inspired so many of the ideas presented in this book. How can one thank Gardner without also saluting Donald E. Knuth? I personally treasure his *Fundamental Algorithms.*[4] Tobias Dantzig's *Number, The Language of Science,*[5] was a true revelation. Indeed, as Einstein himself said, "*This is beyond doubt the most interesting book on the evolution of mathematics which has ever fallen into my hands.*" In D'Arcy Thompson's *On Growth and Form*[6] I encountered the word *gnomon* for the first time, amid an extraordinary display of scholarship and humility, and rare literary form. Ian Stewart

[4] Donald F. Knuth, *Fundamental Algorithms* (Reading, Mass.: Addison-Wesley, 1969).

[5] Tobias Dantzig, *Number: The Language of Science* (New York: Doubleday, 1954).

[6] D'Arcy Wentworth Thompson, *On Growth and Form,* abr. ed. by John Tyler Bonner (Cambridge: Cambridge University Press, 1966).

is continuing the Gardner tradition with his own distinctive talent and, like his predecessor, has the rare gift of making the most difficult subject look simple. John Horton Conway is responsible for many of my sleepless nights, spent inventing little beings endowed with life. I have often used his game to illustrate my lectures on von Neuman's cellular automata. Ah! Roger Penrose's irregular tilings! Eli Maor's *To Infinity and Beyond*[7] and *e: The Story of a Number*[8] were truly illuminating. I wish to pay a particular tribute to George Ifrah's *Histoire universelle des chiffres*,[9] his truly encyclopedic magnum opus, and to Professor Fathy Saleh of Cairo University, who introduced me to ancient Egyptian mathematics. I also wish to express my gratitude to my good friend Arno Penzias, who encouraged me to follow his example and write my own book.[10] Finally, I wish to thank my editor, Trevor Lipscombe at Princeton University Press, for his support and invaluable comments.

[7] Eli Maor, *To Infinity and Beyond* (Princeton: Princeton University Press, 1991).

[8] Eli Maor, *e: The Story of a Number* (Princeton: Princeton University Press, 1994).

[9] George Ifrah, *Histoire universelle des chiffres* (Paris: Editions Robert Laffont, 1994).

[10] Arno Penzias, *Ideas and Information* (New York: W. W. Norton, 1989).

GNOMON

Gnomons

Among Bernouilli's many discoveries, and perhaps the finest
of them all, is the equiangular spiral. It is a curve to be
found in the tracery of the spider's web, in the shells upon
the shore and in the convolutions of the far-away nebulae.
(H. W. Turnbull)[1]

Are spiral galaxies logarithmic? Short of providing a definitive
answer to that question—which answer will need to be grounded in
a coherent theory of how galaxies are formed and why they seem to
whirl[2]—suffice it to observe that spiral galaxies constitute the most
gigantic orderly pattern observable by man (see plate 1). We ourselves
inhabit such a whorled configuration of celestial bodies, and we may
view our habitat only edgewise, which accounts for the milky aspect
of the cluster, hence its name.

Not only do we inhabit a spiral shell like a vulgar gasteropod, but
spirals are found all around us here on earth and seem to permeate
almost every form of life. The shell of the *Nautilus pompilius* (plate 2)
starts with a microscopic seed, to which accretions of material are
added through the years, forming successive chambers whose size
progressively increases as its living inhabitant grows in size. No matter
how large it grows, the shell preserves its original seed, along with
every successive layer. Although neither the sheep's horn nor the beaver's tooth or the tiger's claw give shelter to any living creature, they
grow in the same fashion as the nautilus, as newer material is accreted
to the spiral at the base, which remains attached to the body. The
epeira does not wrap its habitat around its body, but it constructs its
web in the form of a logarithmic spiral, to which French entomologist
Jean-Henri Fabre devoted a mathematical appendix in his *La vie de
l'araignée* (The Life of the Spider).

[1] *The World of Mathematics* (New York: Simon and Schuster, 1956), p. 147.

[2] According to David Malin, "the difficulty lies in that the inclination of the galaxy
on the sky interferes with the necessary measurements, and the spiral structure is
fragile and easily influenced by gravitational interactions with near neighbours."

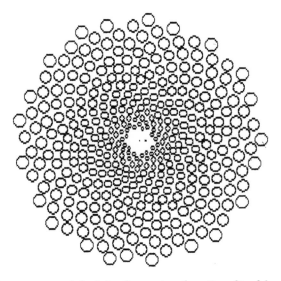

Fig. I.1. Computer model of the flower in plate 3 realized by the author
(34 left-handed and 21 right-handed parastiches).

Logarithmic spirals also govern the arrangement of successive generations of sunflower florets (plate 3 and fig. I.1) cabbage leaves, and
pinecone and pineapple scales, among many natural forms, and it is
not uncommon for petals to wrap themselves around the flower's center,
or leaves around their twig, following beautiful spiral arrangements.
Leonardo de Vinci had already observed that the angle between an
emerging leaf and its predecessor, known as *divergence,* is almost always
constant. That phenomenon may also be observed in the arrangement
of the branches of an aloe (plate 4).

D'Arcy Thomson beautifully captured the property of self-similarity
that he observed in life forms: "For it is peculiarly characteristic of the
spiral shell, for instance, that it does not alter as it grows; each increment
is similar to its predecessor, and the whole, after every spurt of growth,
is just like what it was before."[3]

In all of these constructions, each successive increment is said to
constitute a *gnomon* to the entire structure.

Louis Bravais, a French nineteenth-century botanist, and his brother
Auguste, a physicist, discovered that the divergence in many plant
species approaches 360° divided by 2.618 . . . , the square of the golden

[3] *On Growth and Form,* ed. John Tyler Bonner (Cambridge: Cambridge University
Press, 1966), p. 179.

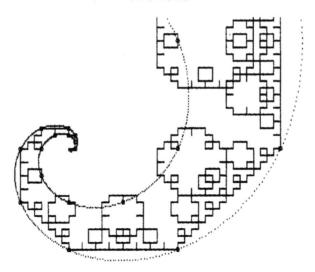

Fig. I.2. A regular fractal with its logarithmic spiral envelope.

number, in which case the number of clockwise and anticlockwise spirals formed by the scales, or *parastiches,* are consecutive Fibonacci numbers that depend upon the speed with which the scales succeed one another. The great Goethe himself, who, unbeknownst to many, was an outstanding naturalist, observed in 1790, "We are now able to study accurately the successive formation of the leaves, since the progressive operations of Nature all take place, step by step, under our eyes."[4]

The overabundance of spirals and self-similar figures in nature has inspired man, since the dawn of humanity, to stylize these figures and utilize them as architectural and decorative motifs, as attested by Prisse d'Avennes's reproductions of ancient Egyptian interleaving spirals in plate 5, as well as the extraordinarily intricate patterns born of the genius of Maurits Escher (plate 6).

Fractals, those figures recently discovered by another Swiss mathematician, Benoît Mandelbrot, constitute a marvelous illustration of self-similarity (fig. I.2 and plate 7). The logarithmic spiral epitomizes *self-similarity*: If you draw such a spiral and photographically enlarge it, the resulting figure is found to be identical to the original, albeit rotated

[4] Johann Wolfgang von Goethe, *Versuch die Metamorphose der Pflanzen zu erklären* (An Attempt to Interpret Plant Metamorphosis). Cited in John E. Dale, "Power Plants," *The Sciences* (October 1994), p. 28.

by an angle that depends on the degree of enlargement. That unique characteristic so fascinated Jakob Bernoulli, the seventeenth-century Swiss mathematician, that he referred to the figure as the *spira mirabilis*, the miraculous spiral, and his last request was to have it engraved on his tombstone, along with the epitaph *Eadem mutata resurgo* (Although changed, I shall arise the same).

The theme of this book is self-similarity, which I call *gnomonicity*, with particular emphasis on the logarithmic spiral and its generating gnomons. The connotation of the term *gnomon* is that originally given by Hero of Alexandria, namely, "A Gnomon is that form that, when added to some form, results in a new form similar to the original."

I open the book with a historical perspective, which some readers might, perhaps justifiably, consider out of context. I am nonetheless inspired by nineteenth-century French sociologist Auguste Comte, who declared that "To understand a science is to know its history." I take this opportunity to point out that the gnomon was an Egyptian invention, and dispel a widespread erroneous reading of the history of the Egyptian obelisk.

Following a study of continued fractions and Fibonacci numbers, I discuss whorled figures and their spiral envelopes and close with a rather unusual presentation of fractals, based on an arithmetized version of the *Kronecker product,* to which the reader will be introduced.

OF GNOMONS AND SUNDIALS

According to hieroglyphic texts, the measurement of time was central to the daily life of ancient Egyptians as early as the Old and the Middle Empires. Early archeological finds do not reveal the presence of solar clocks, however, until the Eighteenth Dynasty.[5] The instrument consisted merely of an L-shaped object with a short vertical arm and a long graduated horizontal arm and was rotated in such a manner that the shadow of the vertical arm fell on the graduations of the horizontal arm. That instrument was called *setchat,* or *merkhet,* literally meaning "instrument of knowledge." According to a text discovered in the Cenotaph of Seti I (1294–1279 B.C.), the first and last hours of the day were not measured, as the shadow cast was too long for practical purposes. That problem was later solved by replacing the horizontal arm by an inclined plane, forming an acute angle with the vertical

[5] The Old Empire is dated between 2700 and 2200 B.C. The Middle Empire lasted from 2033 to 1710 B.C. The clocks date to Thutmosis III, who reigned between 1479 and 1425 B.C.

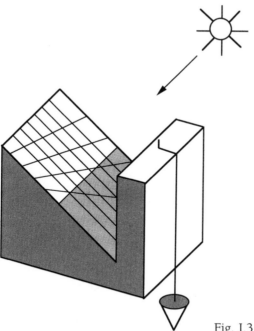

Fig. I.3. An Egyptian sun clock.

needle (fig. I.3). The plane bore six graduated scales, each corresponding to a month, and was symmetrical with respect to the solstices.[6]

The sundial itself did not appear until the Nineteenth Dynasty and consisted of a circle upon which twelve rays were engraved, with a needle at its center. The sundial, unlike the sun clock, needed to be oriented in a fixed direction.

Several centuries later, around 575 B.C., the invention of the sundial was ascribed to the Greek Anaximander, and the term *gnomon,* meaning "that which allows one to know," a literal translation of *merkhet,* came to denote the *L*-shaped object placed at the sundial's center.

The etching in figure I.4 is a 1634 allegory by Girard Desargues (1591–1661), "Universal method of positioning the axle and placing the hours and other items in sundials," and shows the word *Gnomonique* engraved on the sundial held on the Greek figure's lap. Desargues was the founder of projective geometry, and the author of a beautiful theorem in that field.

---

[6] The Egyptian calendar comprised three seasons of four months each, and a month consisted of thirty days. As the year contained 365 days, the remaining five epagomenal days were inserted just before the first day of the year, coinciding with our July 19, which occurred when Sothis, or Sirius, rose with the sun on the horizon. An additional day was added to account for leap years.

7

Fig. I.4. Allegory of Girard Desargues's *Gnomonique*.
Bibliotheque Nationale, Paris.

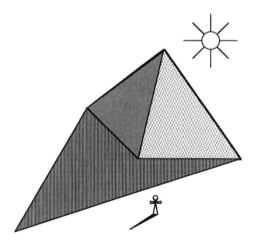

Fig. I.5. Thales measuring the pyramid's height.

On Geometric Similarity

The Greeks were familiar with the notion of geometric similarity since the days of Thales of Miletus, who is given credit by Plutarch for calculating the height of the great pyramid of Giza (fig. I.5). He measured the length of the shadow cast on the ground by a vertical stick and calculated its ratio to the stick's length. At precisely the same time of day, he measured the length of the pyramid's shadow, and by applying the properties of similar triangles he was able to calculate the pyramid's height.

Thales of Miletus, a contemporary of Anaximander, extensively traveled to Egypt, where he studied with the priests, as well as to Mesopotamia. He is rightfully regarded as the founder of Greek geometry, the precursor of all Greek science.[7]

---

[7] Whereas Egyptian and Mesopotamian sciences are usually regarded as essentially applied, and consisting mostly of empirical rules, Thales attempted, following an *ascending* inductive process based on observation, to derive abstract general principles, from which a coherent body of corollary knowledge could then be derived, following a thorough and unforgiving *descending* process of deduction, based on pure reason. That vision is not, however, entirely fair to Egypt and Mesopotamia. It has recently been established—as attested by the *Plimpton* tablet, for example—that Pythagoras, a younger contemporary of Thales, learned in Mesopotamia the famous *theorem* ascribed to his name. Be that as it may, the Greeks, who stood on the shoulders of Egyptian and Mesopotamian giants, are the founders of what was to become the magnificent formal edifice erected in thirteen monumental volumes, the *Elements*, written by Euclid and his disciples around 300 B.C. in the Egyptian city of Alexandria.

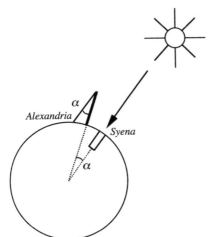

Fig. I.6. Erastothenes calculating the earth's circumference.

Three centuries later, around 245 B.C., Erastothenes, nicknamed Pentathlos (champion in five categories) by his peers, who was made head of the Alexandrian Library by Ptolemaeus, calculated the earth's circumference using the method of triangulation, by measuring the angle α between an obelisk situated in Alexandria and its shadow, precisely as the sun's rays plunged straight down the shaft of a well situated in the city of Syena (Assouan), which was thought to be on the same meridian as Alexandria (fig. I.6). The distance between Syena and Alexandria was then regarded as the fraction of the earth's circumference intercepted by angle α, thus allowing a calculation of the entire circumference,[8] within 7–16% of the presently accepted figure of 40,009 kilometers.

GEOMETRY AND NUMBER

Analogies abound in nature, and innumerable metaphors can be found within one branch of human knowledge for phenomena observed

[8] The invariant relationship between an object and its shadow, be it as colossal as a pyramid or as puny as a stick, inspired science historian Michel Serres to conjecture that the invariance of the *logos*, or ratio, may have led the Greeks to dwell on the insignificance of height and strength, in contrast to the respectability of smallness and humility. A gnomon's shadow is the tangible manifestation on earth, nay, literally on the dust, of celestial events of cosmic magnitude. Perhaps that observation inspired the metaphor of Plato's cave, whose dwellers can perceive only the shadows cast on the cave's walls by unseen perfect objects, reducing them to merely speculate about their intangible reality.

in another. Such is the case with number and geometry, the two corner-stones of Greek science.

The ancient Greeks, who studied the properties of geometric figures as well as those of integers, were obsessed by the search for a unifying theory, whose statements could be spelled out in terms sufficiently abstract to be applicable to numbers as well as geometric figures. To the Greeks, number permeated everything. The world, in its purest and most abstract form, the only one worthy of interest, was built on *commensurate,* or rational, numbers!

What they believed to be a continuum of rational numbers, a gapless succession of numbers expressible as the ratios of two integers, was seen as the perfect metaphor for the similarly gapless geometric figure represented by the line. (That short-lived illusion was shattered by the discovery of the αλογον, the unutterable irrational number, until Eudoxus of Cnidus offered some relief around 360 B.C.).

Two squares were known to be geometrically similar, as were two equilateral triangles, and generally any two regular polygons with the same number of sides. Predicated upon geometric similarity, the notion of number similarity was invented by the Greeks, and they consequently devoted great efforts to the study of figurate numbers: triangular, square, pentagonal, and so on. Whereas the number of beads inside each square box of figure I.7 (top) is to this day referred to as square, other figurate numbers, such as the triangular numbers of figure I.7 (bottom), are no longer accorded any particular merit nowadays, other than being objects of mathematical diversion.

The Greeks had observed that $1 + 3 = 4, 4 + 5 = 9, 9 + 7 = 16, 16 + 9 = 25$, and so on, and obviously knew that the sequence $1, 4, 9, 16, \ldots$ was that of the squares of natural numbers. They were

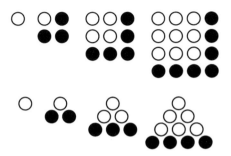

Fig. I.7. *Top,* square numbers. *Bottom,* triangular numbers.

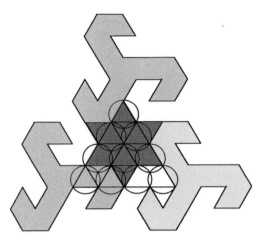

Fig. I.8. The tetraktys, within an Islamic interlocking pattern consisting of stars and a three-armed motif.

struck by the fact that the addition of any term of that sequence to the corresponding term of the sequence 3, 5, 7, 9, . . . of odd numbers resulted in the next term of the squares sequence. Figure I.7a provides a geometrical metaphor for that observation, in which the square numbers are arranged inside a square box, and the odd numbers inside a contiguous *L*-shaped box which, because of its peculiar shape, was called a gnomon. The addition of successive geometric gnomons to a figure does not alter its proportions; it only grows larger, with every new figure similar to its precursors. In like fashion, the addition of successive odd numbers merely extends the squares sequence, without altering its figurate nature. The sequence of square numbers is thus 1, 4, 9, 16, . . . and its gnomonic sequence 3, 5, 7, 9, ... Similarly, the sequence of triangular numbers is 1, 3, 6, 10, 15, . . . and its gnomonic sequence 2, 3, 4, 5, ...

The figure representing 10, the fourth triangular number, was called a *tetraktys* and deemed so mystical by Pythagoras's followers that they adopted it as the emblem of their secret brotherhood, the Order of the Pythagoreans. It is also extensively used in Islamic design, where the tetraktys is used to generate intricate designs involving the six-pointed star and periodic patterns of order 3, such as that of figure I.8.[9]

[9] Many such patterns can be found in Keith Critchlow's beautiful book *Islamic Patterns* (London: Thames and Hudson, 1976).

OF GNOMONS AND OBELISKS

Some modern-day historians contend that the obelisks, several of which were erected in Egypt, were used as sundials. Suffice it to observe the

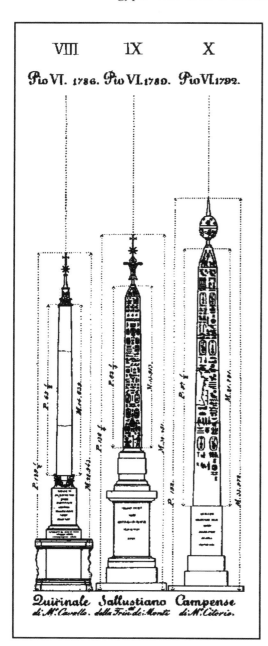

Fig. I.9. Three Egyptian obelisks in Rome. Of particular interest is the obelisk on the right, known as the solar obelisk of Monte Citorio. Not only does it serve as a gnomon, but an opening made in 1788 through the upper globe allows a narrow ray of sunshine to hit an alignment of stones embedded in the pavement, thus allowing the determination of the meridian.

manner in which obelisks often flanked the entrance of temples, as epitomized by the Temple of Luxor where unfortunately only one obelisk remains, while its twin adorns the beautiful Place de la Concorde in Paris (plate 8), to conclude that they were never intended by the Egyptians to serve as sundials, or indeed any purpose other than to glorify the god Amun. According to Egyptologist Labib Habashi, "On the base of her obelisk that still stands in Karnak, (Queen) Hatshepsut recorded that 'she made as her monument for her father Amun, . . . two great obelisks of solid red granite of the region of the south; their upper halves of gold of the best of all countries.' "[10]

The Romans were obviously very fond of Egyptian obelisks, and they transported them across the Mediterranean to adorn the city of Rome, were today one may find no fewer than thirteen standing obelisks, whereas only four remain in Egypt and one in each of Paris, New York, London, and Istanbul, many of them surmounted by crosses or transformed into sundials (fig. I.9).

While it is an established fact that the Egyptians invented the sundial eight centuries before Anaximander, the use of obelisks as gnomons by the ancient Egyptians is no more than an ordinary case of "romanomorphic" reading of history.

[10] Labib Habashi, *The Obelisks of Egypt* (Cairo: American University in Cairo Press, 1988).

Figurate and *m*-adic Numbers

In the introduction, I briefly discussed triangular and square numbers.[1] In this chapter I shall reexamine these numbers as well as some of their successors. I shall also introduce another family of numbers, the *m*-adic numbers, and examine the dyadic and triadic numbers in more detail.

FIGURATE NUMBERS

Plate 9 shows the first four *ranks* of figurate numbers, and for each rank, the first four *orders,* corresponding to $n = 1, 2, 3, 4$, where it is agreed, as a matter of convention, that all figurate numbers are equal to 0 for $n = 0$. A figurate number of rank b and order n is equal to $n + n(n - 1)b/2$. For example, a triangular number of order n is equal to $n(n + 1)/2$, and a square number is equal to n^2. Ever since ancient Greece, generations of mathematicians have been fascinated by figurate numbers and their properties, and Fermat conjectured that any integer could be expressed as the sum of at most m m-gonal numbers. Consider for example the integer 15. We find

Triangular:	$15 = 6 + 6 + 3$
Square:	$15 = 9 + 4 + 1 + 1$
Pentagonal:	$15 = 5 + 5 + 5$
Hexagonal:	$15 = 6 + 6 + 1 + 1 + 1$
Heptagonal:	$15 = 7 + 7 + 1$

In his *Disquisitiones arithmeticae,* Gauss was able to prove that conjecture for triangular and square numbers. It was Cauchy, however, who provided the general proof. When that was achieved, all of Fermat's conjectures had been thoroughly addressed, with the exception of his Last Theorem, which remained unproven until Andrew Wiles only recently put an end to centuries of laborious search.

---

[1] This chapter, more than any other in this book, was inspired by the writings of Martin Gardner, to whom the author is infinitely grateful.

TABLE I.1a
Figurate Numbers and Their Respective Gnomons

Triangular

n	0	1	2	3	4	5	6	7	8
T_n	0	1	3	6	10	15	21	28	36
T_n'	1	2	3	4	5	6	7	8	9
T_n''	1	1	1	1	1	1	1	1	1

Square

n	0	1	2	3	4	5	6	7	8
S_n	0	1	4	9	16	25	36	49	64
S_n'	1	3	5	7	9	11	13	15	17
S_n''	2	2	2	2	2	2	2	2	2

Pentagonal

n	0	1	2	3	4	5	6	7	8
P_n	0	1	5	12	22	35	51	70	92
P_n'	1	4	7	10	13	16	19	22	25
P_n'	3	3	3	3	3	3	3	3	3

Given a figurate number of rank b and order n, what number, when added to it, results in the number of order $(n + 1)$? That is easily found to be

$$\left[(n + 1) + \frac{(n + 1)nb}{2}\right] - \left[n + \frac{n(n - 1)b}{2}\right] = 1 + nb.$$

The figurate number's gnomon is therefore $1 + nb$ in the general case. A trivial family of numbers, those for which $b = 0$, might be called *linear*. The order n linear number is none other than n itself, and its gnomon is 1. The corresponding geometric figure is an n-verticed polygon, all of whose sides are collapsed onto one line. Table 1.1a shows the first few triangular, square, and pentagonal numbers, respectively T_n, S_n, P_n, together with their gnomons T'_n, S'_n, P'_n. An additional line shows T''_n, S''_n, P''_n, the gnomon's gnomon, which is none other than integer b. Clearly, a figurate number of rank n is equal to the sum of its first n gnomons, and in particular *the sum of the first n odd numbers is equal to n^2.*

The reader who is familiar with Newton's finite difference method will remember that, given the values of variable y corresponding to equally spaced values of variable x, a difference table may be constructed, similar to table 1.1b. Putting $A_o = y_0$, Newton's formula, which allows one to express y as a polynomial in x, is

$$ y = \sum_i A_i \binom{x}{i}, $$

where the expression between parentheses represents the number of combinations of x objects taken i at a time, in other words,

$$ \binom{x}{i} = \frac{x!}{i!(x-i)!}. $$

In the case of figurate numbers, the summation is performed over $i = 0, 1, 2, \ldots$. For example, pentagonal numbers correspond to

$$ A_o = 0, \quad A_1 = 1, \quad A_2 = 3 \quad (A_3 = A_4 = A_5 = \cdots = 0); $$

hence, as expected,

$$ P_n = \binom{n}{1} + 3 \binom{n}{2} = n + \frac{3n(n-1)}{2}. $$

A Property of Triangular Numbers

What is the gnomon of a triangular number's square? To answer that question, we turn to figure 1.1. A square is drawn with side $T_n = n(n+1)/2$, to which an L-shaped gnomon is added, as shown. The gnomon's area is equal to $2(n+1)T_n + (n+1)^2 = (n+1)^3$. Figure 1.2 shows a succession of triangular numbers and their squares, highlighting this strange statement: *The square of the sum of the first n integers is equal to the sum of their cubes.*

Nichomachus, the Syrian author of *Introductio Arithmetica*, was a

TABLE 1.1b
Newton's Finite Difference Table

x	0	1	2	3	4
y	y_0	y_1	y_2	y_3	$\vdots$
	$A_1 = y_1 - y_0$	$B_1 = y_2 - y_1$	$C_1 = y_3 - y_2$	$D_1 = y_4 - y_3$	
	$A_2 = B_1 - A_1$	$B_2 = C_1 - B_1$	$C_2 = D_1 - C_1$	$\vdots$	
	$A_3 = B_2 - A_2$	$B_3 = C_2 - B_2$	$C_3 = D_2 - C_2$		
	$\vdots$	$\vdots$			$\vdots$

Plate 1. The galaxy M83. Photograph by David Malin of the Anglo-Australian Observatory, Epping, Australia, by special permission.

Plate 2. The *Nautilus pompilius*. Photograph by the author.

Plate 3. An Egyptian dried flower. Photograph by the author.

Plate 4. Aloe plant. Photograph by the author.

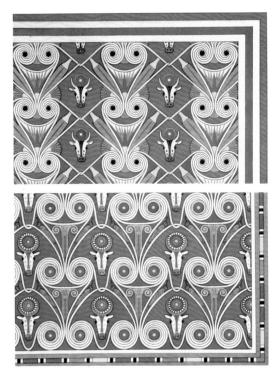

Plate 5. Spiral ornaments in a Seventeenth Dynasty Thebean necropolis, after Prisse d'Avennes.

Plate 6. M. C. Escher, *Whirlpools* (1957). © 1998 Cordon Art B.V.–Baarn–Holland. All rights reserved.

Plate 7. A fractal map showing the character of self-similarity. From Heinz-Otto Peitgen and Peter H. Richter, *The Beauty of Fractals* (Heidelberg: Springer-Verlag, 1986), map 37, p. 82.

Plate 8. The obelisk of Luxor on the Place de la Concorde in Paris, with the recently added golden pyramidion. Photograph by the author.

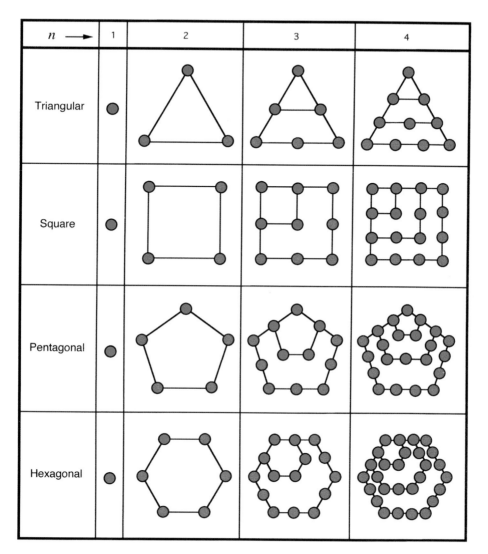

Plate 9. Figurate numbers.

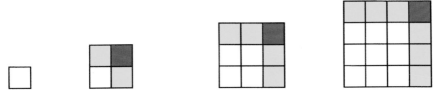

Plate 10. Isometric diagrams of square numbers 1, 4, 9 and their gnomons.

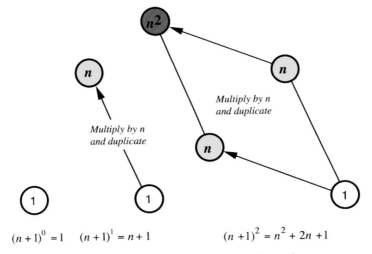

$(n+1)^0 = 1$ $(n+1)^1 = n+1$ $(n+1)^2 = n^2 + 2n + 1$

Plate 11. Construction of the graph of $(n + 1)^2 = n^2 + (2n + 1)$.

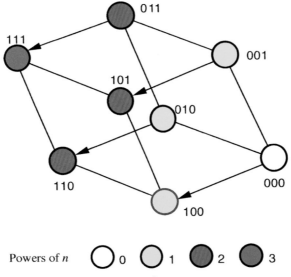

Plate 12. Graph of $(n + 1)^3$, showing vertex indices.

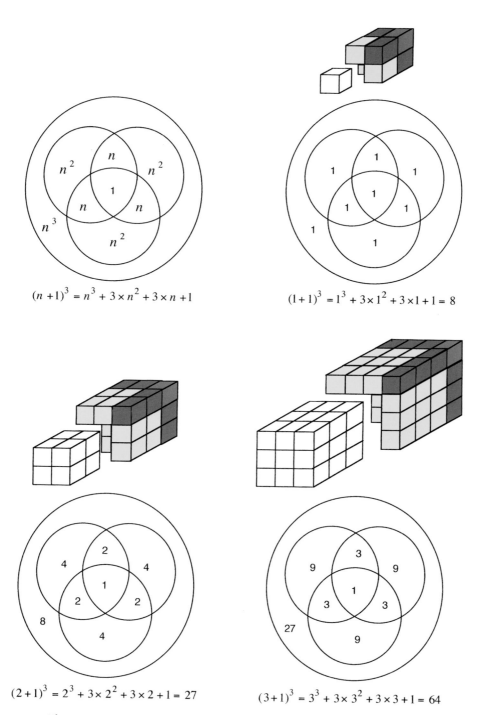

$$(n+1)^3 = n^3 + 3 \times n^2 + 3 \times n + 1$$

$$(1+1)^3 = 1^3 + 3 \times 1^2 + 3 \times 1 + 1 = 8$$

$$(2+1)^3 = 2^3 + 3 \times 2^2 + 3 \times 2 + 1 = 27$$

$$(3+1)^3 = 3^3 + 3 \times 3^2 + 3 \times 3 + 1 = 64$$

Plate 13. Isometric and Venn diagrams of dyadic number cubes and their gnomons.

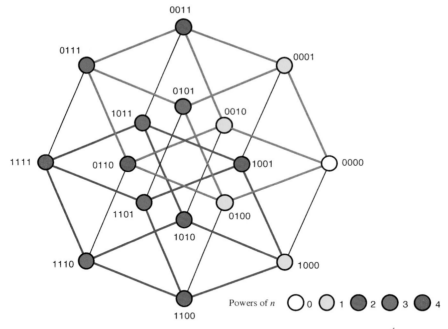

0011
0111
0001
0101
1011
0010
1111 0110 1001 0000
1101 0100
1010
1110 1000
1100

Powers of *n* ◯ 0 ◯ 1 ⬤ 2 ⬤ 3 ⬤ 4

Plate 14. Four-dimensional dyadic hypercubic graph of $(n + 1)^4 = n^4 + 4n^3 + 6n^2 + 4n + 1$.

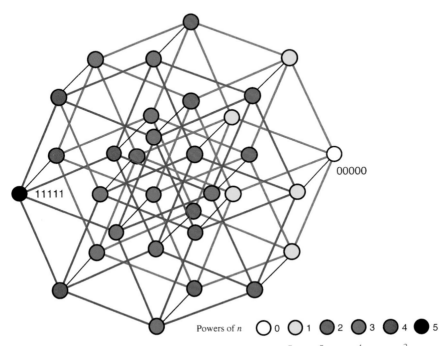

00000

11111

Powers of *n* ◯ 0 ◯ 1 ⬤ 2 ⬤ 3 ⬤ 4 ⬤ 5

Plate 15. Dyadic hypercubic graph of $(n + 1)^5 = n^5 + 4n^4 + 10n^3 + 10n^2 + 4n + 1$.

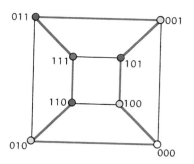

Plate 16. Hamiltonian path over a three-dimensional dyadic cube.

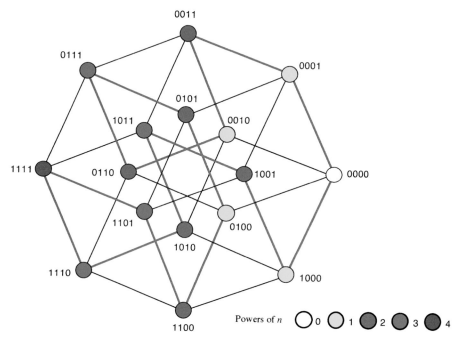

Plate 17. Hamiltonian path over a four-dimensional dyadic hypercube.

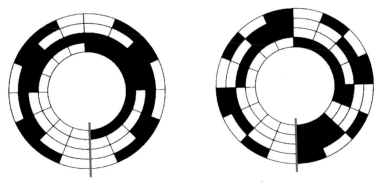

Plate 18. *Left*, Gray encoding disk. *Right*, binary encoding disk.

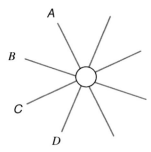

Plate 19. Axes of the hypercube in plate 17.

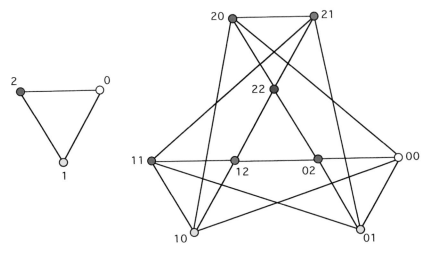

Plate 20. Graphs corresponding to the powers 1 and 2 of triadic numbers.

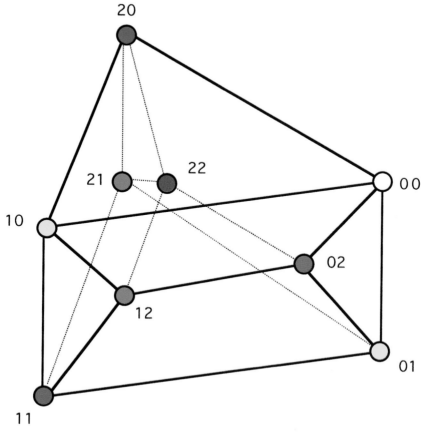

Plate 21a. Isometric view of $(n^2 + n + 1)^2 = n^4 + 2n^3 + 3n^2 + 2n + 1$.

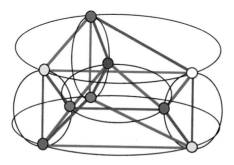

Plate 21b. The four-dimensional triadic hypercube drawn upon a torus, showing two complementary Hamiltonian paths.

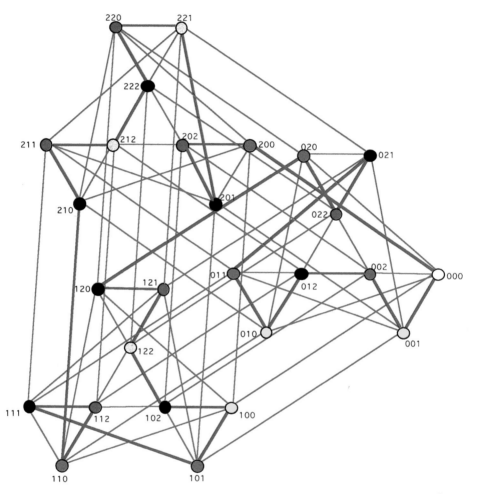

Plate 22. Six-dimensional triadic hypercube corresponding to $(n^2 + n + 1)^3 =$
$n^6 + 3n^5 + 6n^4 + 7n^3 + 6n^2 + 3n + 1$, showing a Hamiltonian path.

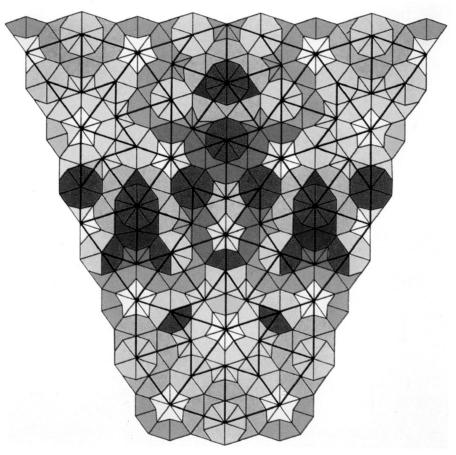

Plate 23. Segment of a Penrose tiled surface. From Martin Gardner, *Penrose Tiles to Trapdoor Ciphers* (New York: W. H. Freeman, 1989), plate 1.

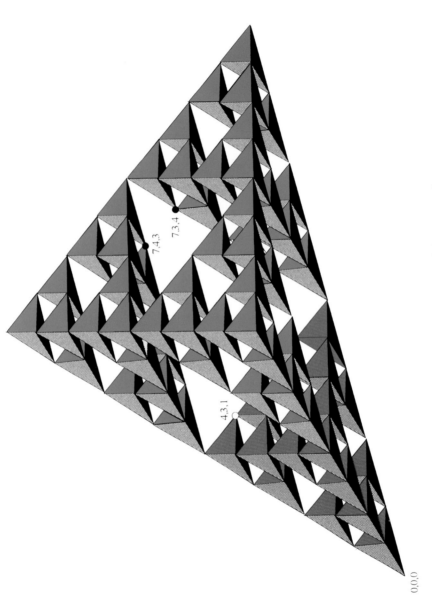

7.4.3

7.3.4

4.3.1

0,0,0

Plate 24. The first three orders of the pyramid.

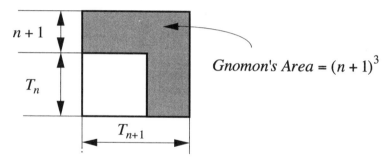

Fig. 1.1. The square of a triangular number.

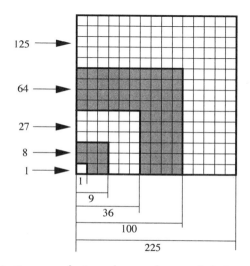

Fig. 1.2. Squares of triangular numbers and their gnomons.

TABLE 1.1c
The Theorem of Nichomachus

n	Sum of the Next n Consecutive Odd Numbers		n^3
1	1	$= 0 \times 1 + (1) \qquad =$	$1^3 = 1$
2	$3 + 5$	$= 2 \times 2 + (1 + 3) \qquad =$	$2^3 = 8$
3	$7 + 9 + 11$	$= 6 \times 3 + (1 + 3 + 5) \qquad =$	$3^3 = 27$
4	$13 + 15 + 17 + 19 = 12 \times 4 + (1 + 3 + 5 + 7) =$		$4^3 = 64$
...	...		...

neo-Pythagorean who lived near Jerusalem, ca. 100 A.D. He made the observation illustrated in table 1.1c, which is closely related to the previous statement. Indeed, the sum of the odd numbers facing integer n in table 1.1c is equal to $n \times (2 \times$ sum of the first $n - 1$ numbers) + sum of the first n odd numbers $= 2nT_{n-1} + n^2 = (n^3 - n^2) + n^2 = n^3$.

A Property of Square Numbers

Plate 10 illustrates the manner in which a square number evolves into the next square number, following the addition of its L-shaped gnomon. In order to transform a square of area $n \times n$ to one of area $(n + 1) \times (n + 1)$, two rectangles of area $n \times 1$ are added onto its sides, as well as a square of area 1×1. That is the geometrical equivalent of the algebraic statement $(n + 1)^2 = n^2 + (2n + 1)$. Another geometric metaphor of that statement, which might be called a *Venn diagram*, shows the relationships between the terms 1, n, n^2 (fig. 1.3). As in plate 10, n^2 is contiguous to n, but not to 1, whereas n is contiguous to n^2 and 1, and 1 is contiguous only to n. In other words, any given power of n is contiguous to those powers from which it differs by 1, and only to those. Figure 1.4 is obtained from figure 1.3 by putting $n = 1, 2, 3$.

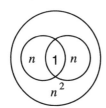

$(n +1)^2 = n^2 + 2n +1$

Fig. 1.3. Venn diagram of $(n + 1)^2 = n^2 + (2n + 1)$.

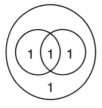   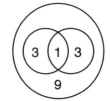

Total = 4   Total = 4 + 2x2 + 1 = 9   Total = 9 + 2x3 + 1 = 16

Fig. 1.4. Venn diagrams of square numbers for $n = 1, 2, 3$.

20

In addition to isometric and Venn diagrams, plate 11 shows yet another metaphor, in the form of a graph in which straight lines connect contiguous powers of n. The latter are shown inside the little bubbles, corresponding to the powers 0, 1, and 2. Going from the graph of $(n + 1)^1$ to that of $(n + 1)^2$ consists of creating an image of the first and sliding it parallel to itself while increasing by 1 the power of n inside each bubble, then connecting it to its image. The graph of $(n + 1)^1$ was obtained from that of $(n + 1)^0 = 1$ in similar fashion. The two sliding operations are done along any two directions upon the plane, provided that they are different.

m-ADIC NUMBERS

We define the m-adic number of order n as the integer

$$(n^{m-1} + n^{m-2} + n^{m-3} + \cdots + n + 1).$$

There exists a single monadic number ($m = 1$), namely, 1. Dyadic numbers are of the form $(n + 1)$, triadic numbers are of the form $(n^2 + n + 1)$, and so on. Table 1.2 shows the first five m-adic numbers, as well as their respective gnomons, for $n = 0$ to 5. The formulae for the gnomons of order n are given in table 1.3. The coefficients

TABLE 1.2
m-adic Numbers and Their Gnomons

n	0	1	2	3	4	5
Monadic No.	1	1	1	1	1	1
Gnomon	0	0	0	0	0	0
Dyadic No.	1	2	3	4	5	6
Gnomon	1	1	1	1	1	1
Triadic No.	1	3	7	13	21	31
Gnomon	2	4	6	8	10	12
Tetradic No.	1	4	15	40	85	156
Gnomon	3	11	25	45	71	101
Pentadic No.	1	5	31	121	341	781
Gnomon	4	26	90	220	440	774

TABLE 1.3
Formulae for the *m*-adic Number Gnomons

Monadic	0
Dyadic	1
Triadic	$2 + 2n$
Tetradic	$3 + 5n + 3n^2$
Pentadic	$4 + 9n + 9n^2 + 4n^3$
Hexadic	$5 + 14n + 19n^2 + 14n^3 + 5n^4$

corresponding to the successive powers of n are shown in figure 1.5a, which, to a certain degree, is reminiscent of Pascal's triangle, and whose structure may be analyzed by the reader as an exercise.

Powers of Dyadic Numbers

The figurate square number of order n is none other than the square of the dyadic number $(1 + n)$. Plate 11 showed how a graph may be transformed from $(n + 1)^0$ to $(n + 1)^1$ and $(n + 1)^2$. Plate 12 shows the transition from $(n + 1)^2$ to $(n + 1)^3$, which consists of sliding the square parallel to itself along a third axis materializing the third dimension, while multiplying the content of each bubble by n, then joining facing bubbles. The result is a *dyadic cube* comprising one bubble containing 1, three-bubbles containing n, three bubbles containing n^2, and one bubble containing n^3. That procedure corresponds to the statement

$$(n + 1)^3 = (n + 1)^2 + n(n + 1)^2 = n^3 + 3n^2 + 3n + 1.$$

The indices are obtained by changing the digit of rank i from 0 to 1 upon duplicating the graph of dimension i − 1 and sliding it parallel to itself. The reader will observe that the power of n within any particular bubble is equal to the number of ones in the corresponding index. Plate 13 shows the isometric as well as the Venn diagrams for $(n + 1)^3$. Again, observe how a power of n is adjacent only to those from which it differs by 1. Plate 14 shows how a *four-dimensional dyadic hypercube* is generated by sliding the three-dimensional cube along a line materializing the fourth dimension. This may be viewed as a projec-

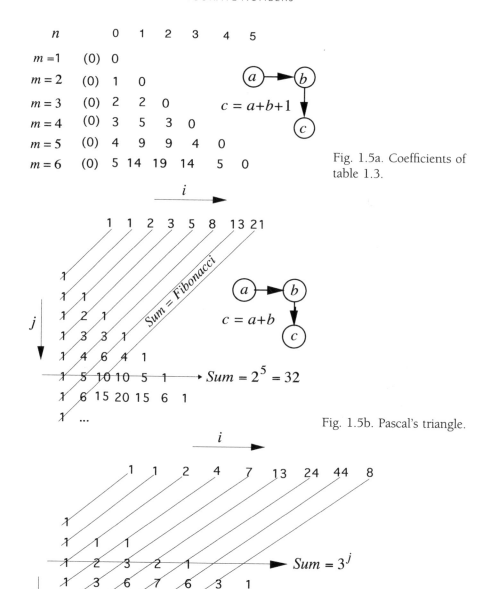

n		0	1	2	3	4	5
$m = 1$	(0)	0					
$m = 2$	(0)	1	0				
$m = 3$	(0)	2	2	0			
$m = 4$	(0)	3	5	3	0		
$m = 5$	(0)	4	9	9	4	0	
$m = 6$	(0)	5	14	19	14	5	0

$c = a+b+1$

Fig. 1.5a. Coefficients of table 1.3.

i

1 1 2 3 5 8 13 21

1
1 1
1 2 1
1 3 3 1
1 4 6 4 1
1 5 10 10 5 1   → $Sum = 2^5 = 32$
1 6 15 20 15 6 1
1 ...

j

Sum = Fibonacci

$c = a+b$

Fig. 1.5b. Pascal's triangle.

i

1 1 2 4 7 13 24 44 8

1
1 1
1 2 3 2 1   → $Sum = 3^j$
1 3 6 7 6 3 1
1 4 10 16 19 16 10 4 1
1 5 15 30 45 ...
1 6 21 ...
1 7 ...
1 ...

j

$d = a+b+c$

Fig. 1.5c. A variation on Pascal's triangle for triadic numbers.

tion of the four-dimensional hypercube upon the plane. The color code corresponds to the powers of n in the expansion $(n + 1)^4 = n^4 + 4n^3 + 6n^2 + 4n + 1$. Again, note that the power of n within any particular bubble is equal to the number of ones, or *weight*, of the corresponding index. Figure 1.6 is a two-dimensional image of the projection of the dyadic four-dimensional hypercube upon three-dimensional space. Plate 15 represents the plane projection of a dyadic five-dimensional hypercube, corresponding to $(n + 1)^5 = n^5 + 4n^4 + 10n^3 + 10n^2 + 4n + 1$.

The previous exercise can be carried on endlessly, and modern-day computers will have no difficulty keeping track of complex interconnections between vertices, which become more and more difficult to visualize. Obviously, the coefficients in the above identity are none other than the *binomial coefficients*, which may be generated following the construction rule of the well-known *Pascal's triangle* (fig. 1.5b):

$$(n + 1)^x = \binom{x}{0} n^x + \binom{x}{1} n^{x-1} + \binom{x}{2} n^{x-2}$$

$$+ \cdots + \binom{x}{j} n^{x-j} + \cdots + \binom{x}{x} n^0.$$

Observe that the sum of the entries on any line x is equal to 2^x, the number of vertices of an x-dimensional dyadic hypercube, and the sum on diagonal $x = j$ is equal to the Fibonacci number $F_{1,x}$.

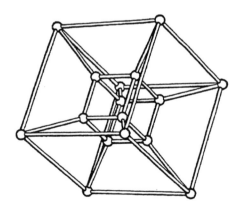

Fig. 1.6. Central projection of the four-dimensional dyadic hypercube.

The Dyadic Hamiltonian Path

Around the middle of the nineteenth century, William Rowan Hamilton, the famous Irish mathematician, invented a game that consisted in following the edges of a regular polyhedron in such a way that all vertices are visited once and only once, in a never-ending cycle. He devoted particular attention to the dodecahedron (10 faces) and icosahedron (20 faces). The latter polyhedron gave its name to the *Icosian game*, which, surprisingly, was sold by Hamilton for the sum of twenty-five pounds! That game can be played on the three-dimensional dyadic cube whose central projection is shown in plate 16, which shows a *Hamiltonian path*. You may wish to enumerate every possible such path. The problem becomes a little more complicated if you are required to perform the same feat on the four-dimensional dyadic hypercube in plate 14.

One possible Hamiltonian path is shown in plate 17. Do not try your hand at counting how many such paths exist, unless you are equipped with a powerful computer.[2] As you travel along the path of plate 17, starting from 0000 toward 0001, you may decide that 0000 is a coded representation of integer 0, 0001 that of integer 1, 0011 that of integer 2, and so on. What you obtain is a perfectly legitimate four-binary digit (or bit) code for the sixteen integers 0 to 15. The correspondence between that code, which is referred to as a binary reflected, or Gray code, and its digital as well as classical binary (or dyadic) equivalents,[3] is shown in table 1.4. You may observe that the binary codes for 7 and 8 differ in all four positions, and that the codes for several other integers vary in more than one position from that integer to the next. On the other hand, any two consecutive Gray codes vary in exactly one bit position, and if a loop is formed of the sixteen integers, whereby integer 0 is the successor of integer 15, the codes corresponding to these two integers will also vary in a single position.

Imagine that encoding disks similar to those of plate 18 are made of transparent material, and that an array of four minute light emitters is placed upon each cursor, one emitter per ring. On the opposite side of the disks, an array of four photosensitive devices is placed facing

[2] See "Gray Codes and Paths on the *n*-cube," *Bell Systems Technical Journal* 37, no. 1 (May 1958), pp. 815–826.

[3] Frank Gray, a research scientist at Bell Labs, filed patent no. 2632058 on March 17, 1953, for the Gray code encoding vacuum tube.

TABLE 1.4

Decimal, Gray, and Binary or Dyadic Codes

Dec.	Gray	Dyadic
0	0 0 0 0	0 0 0 0
1	0 0 0 1	0 0 0 1
2	0 0 1 1	0 0 1 0
3	0 0 1 0	0 0 1 1
4	0 1 1 0	0 1 0 0
5	0 1 1 1	0 1 0 1
6	0 1 0 1	0 1 1 0
7	0 1 0 0	0 1 1 1
8	1 1 0 0	1 0 0 0
9	1 1 0 1	1 0 0 1
10	1 1 1 1	1 0 1 0
11	1 1 1 0	1 0 1 1
12	1 0 1 0	1 1 0 0
13	1 0 1 1	1 1 0 1
14	1 0 0 1	1 1 1 0
15	1 0 0 0	1 1 1 1

the light emitter array. The devices thus constructed may now serve as shaft rotation encoders, with a resolution of $2\pi/16$ radians. Whereas the transition from one code to the next generates a single-bit change in the case of the Gray disk, the binary disk may generate large numbers of spurious intermediate codes, as the light emitters and detectors may never be perfectly aligned. The design of digital mechanical displacement transducers such as shaft rotation sensors, as well as that of early telegraphic transmission apparata, pulse code modulation systems, and the like, led to the development of codes that required that the transition from one value of the signal to the neighboring value did not generate intermediate values. Emile Baudot (1845–1903), a French engineer, invented the *cyclic permuted code,* which is recognized as the precursor of the Gray code. It constitutes the essential idea behind the legendary

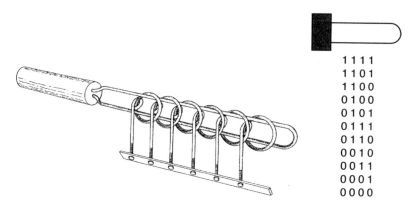

1 1 1 1
1 1 0 1
1 1 0 0
0 1 0 0
0 1 0 1
0 1 1 1
0 1 1 0
0 0 1 0
0 0 1 1
0 0 0 1
0 0 0 0

Fig. 1.7. The baguenaudier. From Martin Gardner, "The Curious Properties of the Gray Code and How It Can Be Used to Solve Puzzles," *Scientific American* (August 1972), p. 66.

success of the Baudot telegraph. Converting from binary to Gray is easily done by replacing the rank i bit by the modulo 2 sum of the ranks i and $i + 1$ bits of the binary number. Conversion from Gray to binary is just as easy; the bit of rank i is replaced by the modulo 2 sum of that bit and every bit to its left.

I shall now describe two famous toys, the *baguenaudier*,[4] which is referred to as the *Chinese rings* in English, and the *Tower of Hanoi*, a game invented by the French mathematician Edouard Lucas, a prolific author of mathematical amusements as well as serious mathematical results. Both games may be analyzed in light of the Gray code.

The baguenaudier is the ring-and-wire puzzle of figure 1.7. The object of the game is to separate the ring chain from the wire loop. The puzzle's structure is such that a ring may come off only if its immediate right-hand neighbor (according to the disposition of the figure) is on the loop, and all the other rings beyond that neighbor are off. According to Martin Gardner, the puzzle was studied by Girolamo Cardano in 1550 in his *De Subtilitate Rerum*, and by John Wallis in 1693 in his *Algebra*. In 1872 Louis Gros applied the binary notation to the puzzle's solution in his *Théorie du Baguenodier*. In the table of figure 1.7, where the number of rings was reduced to four for simplicity, 1 corresponds to a ring on the loop, and 0 to an off ring. The process

---

[4] A French word deriving from a plant whose pod is full of air and explodes with a loud noise upon being pressed between two fingers. The verb deriving from that noun has the connotation of idly wandering about.

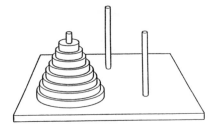

Fig. 1.8. The Tower of Hanoi.

begins with configuration 1111, which corresponds to decimal 10, and the subsequent configurations are made to correspond to 9, 8, 7, . . . , until 0000 is reached.

The tower of Hanoi was invented by Edouard Lucas and sold as a toy in 1883 (fig. 1.8). The object is to transfer the entire tower from one peg to another, removing one disk at a time and threading it in one of the other two pegs, never placing a disk on top of a smaller one.

D. W. Crowe, of the University of British Columbia, discovered that the solution of an n-disk problem is provided by the Hamiltonian path on the n-dimensional dyadic hypercube. To understand how that is done, let us consider the four-disk case, and turn to plate 19 where the axes materializing the dimensions of the hypercube of plate 17 have been named A, B, C, D. If the start node is 0000 and the finish node 1000, and we reduce the Hamiltonian path of figure 1.6 to the enumeration of the axes along which the excursion takes place, ignoring the direction's path upon any one of the axes, that enumeration is *ABACABADABACABA*. Confronted with a four-disk tower, we label the disks A, B, C, D, starting at the top. The puzzle is solved by moving the disks according to the fifteen-letter sequence just described, while respecting the rule that no disk ever be placed on top of a larger one.

That solution therefore involves $2^n - 1$ steps, and its apparent simplicity led to the legend of the "Tower of Brahma," which consisted of sixty-four gold disks. Anyone challenged to transfer it to another location, following the game's rule, may have been unaware that the task would require millions of years to achieve. That legend is reminiscent of the Arabian mathematician who, as a reward for some achievement, required from the caliph that he be granted the quantity of rice obtained by placing one grain on a chessboard square, 2 on the next square, and so on, doubling the amount with every step.

Powers of Triadic Numbers

Whereas the geometric figure corresponding to dyadic $(n + 1)$ is unidimensional, that corresponding to $(n^2 + n + 1)$ is two-dimensional, as shown in plate 20, which also illustrates the manner in which the vertices may be indexed. Plate 21a reveals the existence of six triangular and six parallelipedic faces, and how the figure may be drawn on the

Dec.	Gray	Triadic
0	0 0 0	0 0 0
1	0 0 1	0 0 1
2	0 0 2	0 0 2
3	0 1 2	0 1 0
4	0 1 0	0 1 1
5	0 1 1	0 1 2
6	0 2 1	0 2 0
7	0 2 2	0 2 1
8	0 2 0	0 2 2
9	1 2 0	1 0 0
10	1 2 1	1 0 1
11	1 2 2	1 0 2
12	1 0 2	1 1 0
13	1 0 0	1 1 1
14	1 0 1	1 1 2
15	1 1 1	1 2 0
16	1 1 2	1 2 1
17	1 1 0	1 2 2
18	2 1 0	2 0 0
19	2 1 1	2 0 1
20	2 1 2	2 0 2
21	2 2 2	2 1 0
22	2 2 0	2 1 1
23	2 2 1	2 1 2
24	2 0 1	2 2 0
25	2 0 2	2 2 1
26	2 0 0	2 2 2

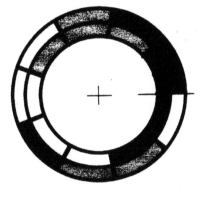

Fig. 1.10. A triadic Gray code and encoding wheel.

29

surface of a torus (plate 21b). It also shows a pair of complementary Hamiltonian paths. Plate 22 shows the six-dimensional triadic hypercube corresponding to

$$(n^2 + n + 1)^3 = n^6 + 3n^5 + 6n^4 + 7n^3 + 6n^2 + 3n + 1,$$

along with two complementary Hamiltonian paths. Whereas Pascal's triangle applies in the dyadic case, the coefficient of n^j in the expansion of $(n^2 + n + 1)^x$ is given in figure 1.5c. Observe that the sum of the entries on any line j is equal to 3^j, the number of vertices of the corresponding triadic hypercube, and that any integer in the top sequence is equal to the sum of the preceding three. A triadic Gray-like code may be constructed, following the vertices on a Hamiltonian path, such as the code shown in figure 1.10. If g_i is the digit of rank i in the Gray code, and t_i the corresponding ternary digit, g_i is given by $g_i = (t_i - t_{i+1} \bmod 3)$. Conversely, t_i is equal to the modulo 3 sum of g_i and all the digits on its left. Figure 1.10 also shows the construction of an encoding wheel whose resolution is 40°. The white, gray, and black colors represent any permutation of integers 0, 1, 2.

Continued Fractions

> Continued fractions are part of the "lost mathematics," the
> mathematics now considered too advanced for high school
> and too elementary for college.
> *(Petr Beckman)*[1]

This chapter is the key to understanding subsequent chapters, which deal with iterative processes such as ladder constructions, Fibonacci sequences, whorled figures, and spirals. It dwells on a special kind of fraction that is widely used to calculate the values of algebraic irrationals such as surds, as well as certain transcendental irrationals, such as e or π. It is believed that continued fractions were first introduced by William Brouncker, the first president of the British Royal Society (1620–1684), who discovered a beautiful expression for the transcendental number π, to which we shall soon return. By merely looking at a continued fraction, one is struck by its iterative character, which appears at once. Whereas the positional representation of quadratic irrationals with respect to a periodic base, be it decimal or otherwise, is not periodic, the corresponding continued fraction is periodic and may thus be defined in terms of a finite number of elements. Similarly, the continued fractions representing e and π obey simple, albeit nonperiodic, patterns. These patterns are epitomized by Euler's continued fractions for e. Whorled figures are the precursors of spirals and constitute perfect geometric metaphors for continued fractions, which accounts for the latter's central role in the chapters to come. As an introduction to continued fractions, we shall first examine a celebrated algorithm dating back to ancient Greece.

EUCLID'S ALGORITHM

The algorithm is found in book 7 of Euclid's *Elements* and is intended to calculate the greatest common divisor of any two integers. Although

---

[1] *A History of π* (New York: St. Martin's Press, 1971), p. 129.

it is ascribed by several historians to the great Eudoxus, it is usually referred to as Euclid's algorithm.

Consider integers a, b with $b > 0$. It is easily established that there exists one and only one integer pair q, r such that

$$a = bq + r, \quad b > r \geq 0. \tag{2.1}$$

Let d be an integer that divides both a and b, and write

$$\alpha = \frac{a}{d}, \quad \beta = \frac{b}{d}. \tag{2.2a}$$

Substituting in equation (2.1), we get

$$\alpha = \beta q + \frac{r}{d}. \tag{2.2b}$$

In the above equations, both α and βq are integers. Consequently, r/d is also an integer, signifying that if d divides both a and b, it also divides r, the remainder of the division of a by b. Conversely, it is clear that any common divisor of b and r is also a divisor of a. The set of common divisors of a and b is therefore identical to the set of common divisors of b and r. The largest number in this set is the greatest common divisor (GCD) of a and b, which is also the GCD of b and r. This is written

$$(a, b) = (b, r). \tag{2.3}$$

That simple property of division is the foundation upon which Euclid based his algorithm, which proceeds as follows.

To determine the GCD of integers a_0 and a_1, where $a_0 > a_1$, we may thus write, following the pattern of equation (2.1),

$$a_0 = a_1 q_0 + a_2, \quad a_1 > a_2$$
$$a_1 = a_2 q_1 + a_3, \quad a_2 > a_3$$
$$a_2 = a_3 q_2 + a_4, \quad a_3 > a_4$$
$$a_3 = a_4 q_3 + a_5, \quad a_4 > a_5$$

$$\cdots$$

$$a_{i-1} = a_i q_{i-1} + a_{i+1}, \quad a_i > a_{i+1}$$
$$a_{n-1} = a_n q_{n-1} + 0. \tag{2.4}$$

Residue $a_n = 0$ is bound to occur eventually for some value n, since integer sequence $a_0 > a_1 > a_2 > a_3 > a_4 \ldots > a_n$ can contain no more than a_0 *positive decreasing* integers. Its occurrence signals that a_n is the GCD of a_{n-1} and itself, that is, $(a_{n-1}, a_n) = a_n$. From the above sequence of divisions, we have, according to equation (2.3),

$$(a_0, a_1) = (a_1, a_2) = (a_2, a_3) = \cdots = (a_{n-1}, a_n).$$

The sought-after GCD is therefore a_n. For example, determine the GCD of 1785 and 374:

$$1785 = 374 \times 4 + 289,$$
$$374 = 289 \times 1 + 85,$$
$$289 = 85 \times 3 + 34,$$
$$85 = 34 \times 2 + 17,$$
$$34 = 17 \times 2 + 0.$$

Hence, $(1785, 374) = 17$.

CONTINUED FRACTIONS

Putting $\phi_0 = \dfrac{a_0}{a_1}$, $\phi_1 = \dfrac{a_1}{a_2}$, $\phi_2 = \dfrac{a_2}{a_3}, \ldots, \phi_{n-1} = \dfrac{a_{n-1}}{a_n}$, equations (2.4) may be written

$$\phi_0 = \frac{a_0}{a_1} = q_0 + \frac{a_2}{a_1} = q_0 + \frac{1}{\phi_1}$$

$$\phi_1 = \frac{a_1}{a_2} = q_1 + \frac{a_3}{a_2} = q_1 + \frac{1}{\phi_2}$$

$$\cdots$$

$$\phi_{n-2} = \frac{a_{n-2}}{a_{n-1}} = q_{n-2} + \frac{a_n}{a_{n-1}} = q_{n-2} + \frac{1}{\phi_{n-1}}$$

$$\phi_{n-1} = \frac{a_{n-1}}{a_n} = q_{n-1} + 0 \tag{2.5}$$

Which in turn can be written

$$\phi_0 = \frac{a_0}{a_1} = q_0 + \cfrac{1}{q_1 + \cfrac{1}{q_2 + \cfrac{1}{q_3 + \cdots}}}$$

$$\cdots q_{n-2} + \cfrac{1}{q_{n-1}} \qquad (2.6a)$$

The above development is called a *continued fraction*, and q_0, q_1, q_2, . . . are called the *partial quotients*. Following the preceding pattern, we may write

$$\phi_0 = \frac{1785}{374} = 4 + \frac{1}{374/289}, \quad q_0 = 4,$$

$$\phi_1 = \frac{374}{289} = 1 + \frac{1}{289/85}, \quad q_1 = 1,$$

$$\phi_2 = \frac{289}{85} = 3 + \frac{1}{85/34}, \quad q_2 = 3,$$

$$\phi_3 = \frac{85}{34} = 2 + \frac{34}{17}, \quad q_3 = 2,$$

$$\phi_4 = \frac{34}{17} = 2, \quad q_4 = 2, \qquad (2.6b)$$

and we get

$$\phi_0 = \frac{1785}{374} = 4 + \cfrac{1}{1 + \cfrac{1}{3 + \cfrac{1}{2 + \cfrac{1}{2}}}}. \qquad (2.6c)$$

SIMPLE CONTINUED FRACTIONS

The previous discussion suggests the following general form for continued fractions:

$$\phi = q_0 + \cfrac{p_1}{q_1 + \cfrac{p_2}{q_2 + \cfrac{p_3}{q_3 + \cdots}}} \qquad (2.7)$$
$$\cdots q_{n-1} + \cfrac{p_n}{q_n},$$

of which striking examples are given by expressions (2.17) and (2.18) below. A particular case results from putting $p_i = 1$ for all i, which corresponds to what is called a *simple continued fraction* (SCF). If, additionally, all q's are positive integers, the SCF is said to be a *regular continued fraction* (RCF). The following notation will be used for simple continued fractions, whether they are regular or not:

$$[q_0, q_1, q_2, \ldots, q_n] = q_0 + \cfrac{1}{q_1 + \cfrac{1}{q_2 + \cfrac{1}{q_3 + \cdots}}} \qquad (2.8)$$
$$\cdots q_{n-1} + \cfrac{1}{q_n},$$

and we can easily show that

$$[q_0, q_1, q_2, \ldots, q_n] = \left[q_0, [q_1, q_2, \ldots, q_{n-1}] \right] = q_0 + \frac{1}{[q_1, q_2, \ldots, q_n]},$$
$$a[q_0, q_1, q_2, \ldots] = \left[aq_0, \frac{q_1}{a}, aq_2, \frac{q_3}{a}, \ldots \right]. \qquad (2.9)$$

CONVERGENTS

The irreducible fraction

$$\delta_i = [q_0, q_1, q_2, \ldots, q_i] = \frac{N_i}{D_i} \qquad (2.10)$$

is called the continued fraction's *ith convergent*. N_i and D_i are the fraction's ith numerator and denominator, respectively. Returning to the numerical example, we get

$$\delta_0 = 4,$$

$$\delta_1 = 4 + \frac{1}{1} = 5,$$

$$\delta_2 = 4 + \cfrac{1}{1 + \cfrac{1}{3}} = \frac{19}{4},$$

$$\delta_3 = 4 + \cfrac{1}{1 + \cfrac{1}{3 + \cfrac{1}{2}}} = \frac{49}{9},$$

$$\delta_4 = 4 + \cfrac{1}{1 + \cfrac{1}{3 + \cfrac{1}{2 + \cfrac{1}{2}}}} = \frac{105}{22} = \frac{1785}{374}.$$

If, for consistency, we introduce "virtual" numerators

$$N_{-1} = 1, \qquad N_{-2} = 0, \qquad D_{-1} = 0, \qquad D_{-2} = 1, \quad (2.11)$$

we may derive two fundamental recursion formulae that, together, yield the successive *irreducible* convergents:

$$N_i = N_{i-2} + q_i N_{i-1} \quad \text{and} \quad D_i = D_{i-2} + q_i D_{i-1}. \quad (2.12)$$

Hence the procedure illustrated in Table 2.1.
    Huygens showed that in the general case,

$$N_{i-1} D_i - N_i D_{i-1} = (-1)^i. \quad (2.13)$$

As an exercise, the reader may also verify that in the case of continued fraction (2.7), putting

$$N_{-1} = 1, \qquad N_{-2} = 0, \qquad D_{-1} = 0, \qquad D_{-2} = 1, \qquad \text{and } p_0 = 1$$

TABLE 2.1
Convergents

i	−2	−1	0	1	2	3	4
q_i	—	—	4	1	3	2	2
N_i	0	1	4	5	19	43	105
D_i	1	0	1	1	4	9	22
∂_i	—	—	4	5	4.75	4.7	4.7$\underline{72}$

yields

$$N_i = p_i N_{i-2} + q_i N_{i-1} \quad \text{and} \quad D_i = p_i D_{i-2} + q_i D_{i-1}, \quad (2.14)$$

where N_i and D_i are the numerator and denominator of convergent ∂_i.

TERMINATING REGULAR CONTINUED FRACTIONS

Let us return to development (2.6b), where the number ϕ_0 may or may not be an integer, and write

$$\phi_0 = q_0 + \frac{1}{\phi_1}, \quad \text{with } \phi_1 > 1,$$

$$\phi_1 = q_1 + \frac{1}{\phi_2}, \quad \text{with } \phi_2 > 1,$$

$$\phi_2 = q_2 + \frac{1}{\phi_3}, \quad \text{with } \phi_3 > 1,$$

$$. \quad . \quad .$$

Index s is reached, for which $\phi_s = q_s$ is an integer. The process halts, and we are in presence of the *terminating RCF*

$$\phi_0 = [q_0, q_1, q_2, \ldots, q_s],$$

signaling that ϕ_0 is a rational number.

Every terminating RCF thus generates a rational number, and, applying Euclid's algorithm, every rational number can be generated by an RCF.

Quotient q_s is called the *terminating quotient* or *termination*.

PERIODIC REGULAR CONTINUED FRACTIONS

Indices s and t are reached, such that $\phi_{s+t} = \phi_s$ and s is the lowest index for which that occurs, resulting in $\phi_{s+a} = \phi_{s+(a \bmod t)}$, for $a = 0, 1, 2, \ldots$. The SCF is said to be periodic, of lead length s, and period t.

Example. To find the RCF corresponding to $\sqrt{14}$, we write[2]

$$
\begin{aligned}
\phi_0 &= \sqrt{14} & &= 3 + (\sqrt{14} - 3), \\
\phi_1 &= \frac{1}{\sqrt{14} - 3} & &= 1 + \frac{\sqrt{14} - 2}{5}, \\
\phi_2 &= \frac{5}{\sqrt{14} - 2} & &= 2 + \frac{\sqrt{14} - 2}{2}, \\
\phi_3 &= \frac{2}{\sqrt{14} - 2} & &= 1 + \frac{\sqrt{14} - 3}{5}, \\
\phi_4 &= \frac{5}{\sqrt{14} - 3} & &= 6 + (\sqrt{14} - 3), \\
\phi_5 &= \frac{1}{\sqrt{14} - 3} & & \quad .
\end{aligned}
\tag{2.15}
$$

In this example, $\phi_5 = \phi_1$, meaning that the resulting RCF is of lead length 1 and period 4. Indeed, $\sqrt{14} = [3, 1, 2, 1, 6, 1, 2, 1, 6, 1, 2, \ldots] = [3, \overline{1, 2, 1, 6}]$. Another easy example is supplied by $\sqrt{2}$:

---

[2] The method used for generating the identities in the example is described in the appendix.

Given that $2\sqrt{2} + 3 = (\sqrt{2} + 1)^2$, we may write

$$\phi_0 = \sqrt{2} + 1 = \frac{2\sqrt{2} + 3}{\sqrt{2} + 1} = 2 + \frac{1}{\sqrt{2} + 1} = 2 + \frac{1}{\phi_0}.$$

Hence,

$$\sqrt{2} + 1 = [2, 2, 2, 2, \ldots]$$
$$\sqrt{2} = [1, 2, 2, 2, \ldots] = [1, \underline{2}].$$

An underlining scheme similar to that used in connection with periodic positional representations can be used, as in the above expressions. Using the tabular method to calculate the successive convergents of $\sqrt{2}$, we get the numbers in table 2.2. The successive convergents oscillate around asymptotic value $\sqrt{2}$:

$$
\begin{array}{ccccccc}
 & 1.5 & & 1.416 & & 1.4142857\ldots & & & \ldots \\
 \uparrow & \downarrow & \uparrow & \downarrow & & \uparrow & & \downarrow & & \uparrow \\
 1 & & 1.4 & & 1.413793\ldots & & & & 1.4142012\ldots
\end{array}
$$

That pattern, which is typical of SCFs, is reminiscent of the damped oscillations so familiar to electrical engineers. In the words of H. W. Turnbull, "As these successive ratios are alternately less than and greater

TABLE 2.2
Convergence to $\sqrt{2}$

i	-2	-1	0	1	2	3	4	5	6
q_i	—	—	1	2	2	2	2	2	2
N_i	0	1	1	3	7	17	41	99	239
D_i	1	0	1	2	5	12	29	70	169
∂_i	—	—	1	$\frac{3}{2}$	$\frac{7}{5}$	$\frac{17}{12}$	$\frac{41}{29}$	$\frac{99}{70}$	$\frac{239}{169}$

TABLE 2.3
Convergence to $\phi = (1 + \sqrt{5})/2$

i	-2	-1	0	1	2	3	4	5	6
q_i	—	—	1	1	1	1	1	1	1
N_i	0	1	1	2	3	5	8	13	21
D_i	1	0	1	1	2	3	5	8	13
∂_i	—	—	1	2	$\dfrac{3}{2}$	$\dfrac{5}{3}$	$\dfrac{8}{5}$	$\dfrac{13}{8}$	$\dfrac{21}{13}$

than all that follow, they nip the elusive limiting ratio between two extremes, like the ends of a closing pair of pincers. They approximate from both sides to the desired irrational. Like pendulum swings of an exhausted clock, they die down—but they never actually come to rest."

The example illustrates the use of continued fractions for the approximation of quadratic irrational numbers. A reasonably good and easy-to-remember aproximation of $\sqrt{2}$ is provided by ∂_5 above, namely, 99/70. That is exactly the proportion of a standard sheet of stationery in France, whose sides measure 29.7 × 21 centimeters, a remarkable choice whose merit will soon become apparent.

A further example is provided by the often cited quadratic number known as the golden section, namely, $\phi = (1 + \sqrt{5})/2 = [1, 1, 1, 1, \ldots]$. Table 2.3 shows its first seven convergents. It may be shown that a periodic RCF always converges to a quadratic algebraic irrational, that is, a number of the form $(a + \sqrt{b})/c$, where a, b, c are integers, and b is not a perfect square. Lagrange proved the converse of that statement, namely, that any quadratic number can be expressed in the form of a periodic SCF.

Spectra of Surds

Early Arabic mathematical textbooks referred to square roots as *samet*, meaning mute, which was transposed to *surd* in the original translations from the Arabic. A surd's continued fraction is always of the form

$$\sqrt{N} = [\alpha, \beta, \chi, \delta, \ldots \delta, \chi, \beta, \omega], \quad \text{where } \omega = 2\alpha,$$

as in $\sqrt{2} = [1, \underline{2}]$, $\sqrt{3} = [1, \underline{1, 2}]$, $\sqrt{5} = [2, \underline{4}]$, $\sqrt{14} = [3, \underline{1, 2, 1, 6}]$.

Tobias Dantzig referred to the integers between square brackets as the surd's *spectrum*. Table 2.4 gives the spectra corresponding to a few surds. The RCFs are periodic, following a *lead sequence* of length 1 consisting of integer α, which is the integral part of the surd. The RCFs are said to be *quasi-periodic*. If the lead sequence is absent, the RCF is *strictly periodic*.

TABLE 2.4
Spectra of Surds

N	α		ω
2	1		2
3	1	1	2
5	2		4
6	2	2	4
7	2	1 1 1	4
8	2	1	4
10	3		6
11	3	3	6
12	3	2	6
13	3	1 1 1 1	6
14	3	1 2 1	6
15	3	1	6
17	4		8
18	4	4	8
19	4	2 1 3 1 2	8
20	4	2	8
21	4	1 1 2 1 1	8
22	4	1 2 4 2 1	8
23	4	1 3 1	8
24	4	1	8

NONPERIODIC NONTERMINATING REGULAR CONTINUED FRACTIONS

The SCF is neither terminated nor periodic, which corresponds to algebraic numbers of degree higher than 2 and to transcendental (non-algebraic) irrationals such as e, for which Euler (1707–1783) discovered the representation

$$e = [2, 1, 2, 1, 1, 4, 1, 1, 6, 1, 1, 8, \ldots]. \tag{2.16}$$

The convergence pattern of Table 2.5 is similar to that observed in the case of periodic continued fractions: *The number is approached by successive ratios that are alternately less and greater than all that follow.* With ten-decimal digit precision, convergent $\partial_6 \approx 2{,}717948718 \ldots$ is only 0.012% removed from e.

It can be shown that every nonterminating RCF is convergent and that every real number can be expanded in exactly one way as an RCF. If we regard a number's expansion into an RCF as a "representation" of that number within the "system" of continued fractions, we are struck by the similarities—and differences—with the positional number system. Both representations are always convergent and uniquely correspond to a number. Conversely, every number can be uniquely represented in either system. In both cases, terminating expansions represent rational numbers. Whereas infinite periodic representations correspond to rational numbers in the positional system, they correspond to quadratic irrationals in the RCF system. Transcendental irrationals are obviously represented by infinite nonperiodic representations in both

TABLE 2.5
Convergence to e

i	-2	-1	0	1	2	3	4	5	6
q_i	—	—	2	1	2	1	1	4	1
N_i	0	1	2	3	8	11	19	87	106
D_i	1	0	1	1	3	4	7	32	39
∂_i	—	—	2	3	$\dfrac{8}{3}$	$\dfrac{11}{4}$	$\dfrac{19}{7}$	$\dfrac{87}{32}$	$\dfrac{106}{39}$

cases. Brouncker discovered the following beautiful expression for the transcendental number π:

$$\frac{4}{\pi} = 1 + \cfrac{1}{2 + \cfrac{9}{2 + \cfrac{25}{2 + \cfrac{49}{2 + \ldots}}}} \tag{2.17}$$

Euler made substantial contributions to the study of continued fractions and discovered a no less beautiful expression for e, the other famous transcendental number:

$$e = 2 + \cfrac{1}{1 + \cfrac{1}{2 + \cfrac{2}{3 + \cfrac{3}{4 + \ldots}}}} \tag{2.18}$$

Continued fractions offer a unique insight into infinite processes such as positional numeration, spirals, and Fibonacci sequences. But let us first introduce the notion of retrovergents.

RETROVERGENTS

Consider the terminating continued fraction

$$\phi = [q_0, q_1, q_2, \ldots, q_{s-1}, q_s].$$

Starting at the far end, we shall write

$$\rho_0 = q_s,$$

$$\rho_1 = [q_{s-1}, q_s] = q_{s-1} + \frac{1}{\rho_0},$$

$$\rho_2 = [q_{s-2}, q_{s-1}, q_s] = q_{s-2} + \frac{1}{\rho_{s-1}},$$

$$\cdots$$

$$\rho_s = \phi.$$

We will refer to $\rho_0, \rho_1, \rho_2, \ldots, \rho_s$ as *retroconvergents,* or *retrovergents.*
Returning to the continued fraction $\phi = [4, 1, 3, 2]$, we have

$$\rho_0 = 2,$$

$$\rho_1 = 2 + \frac{1}{2} = \frac{5}{2},$$

$$\rho_2 = 3 + \frac{2}{5} = \frac{17}{5},$$

$$\rho_3 = 1 + \frac{5}{17} = \frac{22}{17},$$

$$\phi = \rho_4 = 4 + \frac{17}{22} = \frac{105}{22} = \frac{1785}{374}.$$

The significance of this notion will become apparent through the study of *whorled rectangles,* its geometric metaphor.

In later chapters, we shall be dealing with simple continued fractions that are not necessarily regular, such as

$$\sqrt{2} = \left[\frac{1}{\sqrt{2}}, \frac{1}{\sqrt{2}}, \frac{1}{\sqrt{2}}, \ldots, \frac{1}{\sqrt{2}}, \sqrt{2} \right]. \qquad (2.19)$$

The abbreviation SPF will apply to periodic simple continued fractions, regular or not.

APPENDIX

How do we get to the statement

$$\frac{5}{\sqrt{14} - 2} = 2 + \frac{\sqrt{14} - 2}{2}?$$

1. We put

$$x = \frac{5}{\sqrt{14} - 2}.$$

2. Multiplying the numerator and denominator by $\sqrt{14} + 2$, we get

$$x = \frac{5(\sqrt{14} + 2)}{10} = \frac{\sqrt{14} + 2}{2},$$

hence $2x - 2 = \sqrt{14}$, which yields

$$(x - 1)^2 = \frac{14}{4}.$$

We now plug in successive values of $x = 1, 2, 3, \ldots$ until one such value makes the left-hand side larger than the right-hand side:

$$x = 1 \rightarrow (x - 1)^2 = 0,$$
$$x = 2 \rightarrow (x - 1)^2 = 1,$$
$$x = 3 \rightarrow (x - 1)^2 = 4 > \frac{14}{4}.$$

The largest value of x that causes the left-hand side not to exceed the right-hand side is 2 and is the integral part of x.

3. This allows us to write

$$x = \frac{\sqrt{14} + 2}{2} = 2 + y,$$

where $y < 1$ is the fractional part of x. Solving for y, we get

$$y = \frac{\sqrt{14} - 2}{2},$$

and we obtain

$$\frac{5}{\sqrt{14} - 2} = 2 + \frac{\sqrt{14} - 2}{2}.$$

SUMMARY OF FORMULAE

General Form of Continued Fraction (CF)

$$\phi = q_0 + \cfrac{p_1}{q_1 + \cfrac{p_2}{q_2 + \cfrac{p_3}{q_3 + \cdots}}}$$

$$\cdots q_{n-1} + \frac{p_n}{q_n}.$$

45

Examples
Terminating CF (*n* finite):

$$\frac{105}{76} = 1 + \cfrac{1}{2 + \cfrac{9}{2 + \cfrac{25}{2}}}.$$

Nonterminating CF (*n* infinite):

$$\frac{4}{\pi} = 1 + \cfrac{1}{2 + \cfrac{9}{2 + \cfrac{25}{2 + \cfrac{49}{2 + \cdots}}}}.$$

Simple Continued Fractions (SCFs): $p_i = 1$ for all i

$$\phi = [q_0, q_1, q_2, \ldots q_n] = q_0 + \cfrac{1}{q_1 + \cfrac{1}{q_2 + \cfrac{1}{q_3 + \cdots}}}$$

$$\cdots q_{n-1} + \cfrac{1}{q_n},$$

$$[q_0, q_1, q_2, \ldots, q_n] = [q_0, [q_1, q_2, \ldots, q_n]] = q_0 + \frac{1}{[q_1, q_2, \ldots, q_n]},$$

$$a[q_0, q_1, q_2, \ldots] = \left[aq_0, \frac{q_1}{a}, aq_2, \frac{q_3}{a}, \ldots \right].$$

If q_i is a positive integer for all i, the SCF is said to be regular and denoted RCF.

Examples
Terminating RCF (*n* finite): $87/32 = [2, 1, 2, 1, 1, 4]$

Nonterminating RCF (*n* infinite): $e = [2, 1, 2, 1, 1, 4, 1, 1, 6, \ldots]$

Convergents

$$\delta_0 = q_0, \qquad \delta_1 = q_0 + \frac{1}{q_1} = [q_0, q_1], \qquad \delta_2 = [q_0, q_1, q_2], \qquad \text{etc.}$$

Recursive formulae for the convergent:

$$\delta_i = [q_0, q_1, q_2, \ldots q_i] = \frac{N_i}{D_i},$$

$$N_{-1} = 1, \qquad N_{-2} = 0, \qquad D_{-1} = 0, \qquad D_{-2} = 1,$$

$$N_i = N_{i-2} + q_i N_{i-1} \quad \text{and} \quad D_i = D_{i-2} + q_i D_{i-1}.$$

Huygens's formula: $\quad N_{i-1} D_i - N_i D_{i-1} = (-1)^i.$

Retrovergents

$$\phi = [q_0, q_1, q_2, \ldots, q_{s-1}, q_s],$$

$$\rho_0 = q_s,$$

$$\rho_1 = [q_{s-1}, q_s] = q_{s-1} + \frac{1}{\rho_0},$$

$$\rho_2 = [q_{s-2}, q_{s-1}, q_s] = q_{s-2} + \frac{1}{\rho_{s-1}},$$

$$\ldots,$$

$$\rho_s = \phi.$$

Semiperiodic RCF of Lead Length s and Period t

$$\phi = [q_0, q_1, q_2, \ldots q_{s-1}, \alpha, \alpha', \alpha'', \alpha''', \ldots, \alpha^{(t-1)},$$
$$\alpha, \alpha', \alpha'', \alpha''', \ldots, \alpha^{(t-1)}, \alpha, \alpha', \alpha'', \ldots]$$
$$= [q_0, q_1, q_2, \ldots, q_{s-1}, \underline{\alpha, \alpha', \alpha'', \alpha''', \ldots, \alpha^{(t-1)}}].$$

Surds: $\sqrt{N} = [\alpha, \beta, \chi, \delta, \ldots, \delta, \chi, \beta, \omega]$, where $\omega = 2\alpha$,
$\sqrt{2} = [1, \underline{2}]$, $\quad \sqrt{3} = [1, \underline{1, 2}]$, $\quad \sqrt{5} = [2, \underline{4}]$, $\quad \sqrt{14} = [3, \underline{1, 2, 1, 6}]$.

Periodic SCF of Period t

$$\phi = [\alpha, \alpha', \alpha'', \alpha''', \ldots, \alpha^{(t-1)}, \alpha, \alpha', \alpha'', \alpha''', \ldots, \alpha^{(t-1)}, \alpha, \alpha', \alpha'', \ldots]$$

$$= [\underline{\alpha, \alpha', \alpha'', \alpha''', \ldots, \alpha^{(t-1)}}]$$

Fibonacci Sequences

The Fibonacci sequence has intrigued mathematicians for
centuries, partly because it has a way of turning up in
unexpected places but mainly because the veriest amateur
in number theory, with no knowledge beyond simple
arithmetic, can explore the sequence and discover a
seemingly endless variety of curious theorems. Recent
developments in computer programming have reawakened
interest in the series because it turns out to have useful
applications in the sorting of data, information retrieval,
[and] the generation of random numbers.
(Martin Gardner)[1]

Edouard Lucas, a nineteenth-century French mathematician who
published a monumental four-volume book of recreational mathematics, discovered in Leonardo of Pisa's *Liber Abaci* a number sequence
that he himself had studied at length, but for which he nonetheless
gave credit to Leonardo, and which became known as the sequence of
Fibonacci, Leonardo's surname. That sequence has fascinated generations of serious mathematicians as well as tinkerers, and its innumerable
virtues have been found in the arrangement of sunflower seeds, pine-
cones, the divine section, electrical networks, Bernouilli's miraculous
spiral, and other places. New properties are discovered or rediscovered
every day around the globe and could fill several volumes. Fibonacci
fan clubs have even been created, and *The Fibonacci Quarterly* was
launched in 1963 by the Fibonacci Association and edited by Verner
E. Hoggatt Jr. of San Jose State College. Be that as it may, we have
decided to include a section on the sequence in this book because of
its immediate relevance to the study of recursive processes, and in
particular those generating gnomonic, or self-similar figures, such as
geometric fractals and their spiral envelopes. We believe that, in the
process, we may have made one or two modest contributions, though
it is difficult to assess novelty in such a turbulent arena.

[1] *Mathematical Circus* (Middlesex: Penguin Books, 1979), p. 155.

The Fibonacci sequence (FS) classically begins with a pair of successive 1's, and each subsequent integer is equal to the sum of the two integers immediately preceding it, as follows:

$$1, 1, 2, 3, 5, 8, 13, 21, \ldots .$$

If the sequence is extended leftwards, allowing for negative integers, an arrangement is obtained which is infinite in both directions:

$$\ldots -21, 13, -8, 5, -3, 2, -1, 1, 0, 1, 1, 2, 3, 5, 8, 13, 21 \ldots .$$

It is symmetrical around zero, except that every other integer left of zero carries a negative sign. If we construct the sequence of those numbers equal to the difference between two consecutive Fibonacci numbers and refer to it as the Fibonacci gnomonic sequence, we discover that it is identical to the initial sequence, albeit "out of phase" with the latter. We shall say that the Fibonacci sequence is *homognomonic*. Fibonacci sequences have been generalized in many ways, including Lucas numbers, Mark Feinberg's Tribonacci numbers, and so on. We shall restrict our choice to a simple generalization that we have found to be useful with regard to the properties of spirals as well as other interesting mathematical creatures and manmade objects.

RECURSIVE DEFINITION

A Fibonacci sequence of order m, where m is a real positive number, is defined as the sequence whose term $F_{m,n}$ is given by

$$F_{m,0} = 0, \qquad F_{m,1} = 1,$$
$$F_{m,n+2} = F_{m,n} + mF_{m,n+1}. \tag{3.1}$$

As an example, sequences of order 1, 2, and $1/\sqrt{2}$ are given in Table 3.1 for $n = -4$ to 5.

THE SEED AND GNOMONIC NUMBERS

Given m, the order of the Fibonacci sequence, we define Φ_m as follows:

TABLE 3.1
Values of $F_{m,n}$ for $m = 1, 2, 1/\sqrt{2}$ and $n = -4$ to 5

n m	-4	-3	-2	-1	0	1	2	3	4	5
1	-3	2	-1	1	0	1	1	2	3	5
2	-12	5	-2	1	0	1	2	5	12	29
$1/\sqrt{2}$	$-5/(2\sqrt{2})$	$3/2$	$-1/\sqrt{2}$	1	0	1	$1/\sqrt{2}$	$3/2$	$5/(2\sqrt{2})$	$11/4$

$$m = \Phi_m - \frac{1}{\Phi_m}, \quad \text{with } \Phi_m > 1. \tag{3.2}$$

It follows that

$$\Phi_m^2 - m\Phi_m - 1 = 0, \tag{3.3a}$$

$$\Phi_m = \frac{\sqrt{4 + m^2} + m}{2}, \qquad \frac{1}{\Phi_m} = \frac{\sqrt{4 + m^2} - m}{2},$$

$$\Phi_m + \frac{1}{\Phi_m} = \sqrt{4 + m^2}. \tag{3.3b}$$

We can also write

$$\Phi_m = mD_m, \quad \text{with } D_m = \frac{1}{2}\left(1 + \sqrt{1 + \frac{4}{m^2}}\right), \tag{3.3c}$$

$$\Phi_m^2 = \frac{D_m}{D_m - 1}, \qquad D_m = \frac{\Phi_m^2}{\Phi_m^2 - 1}. \tag{3.3d}$$

For reasons that will later on become apparent, we shall refer to Φ_m as the *seed number* of m, and to m as the *gnomonic number* of Φ_m.

Examples

$$\Phi_1 = \frac{\sqrt{5} + 1}{2}, \qquad \Phi_2 = 1 + \sqrt{2}, \qquad \Phi_{1/\sqrt{2}} = \sqrt{2}.$$

EXPLICIT FORMULATION OF $F_{m,n}$

In equations (3.1), $F_{m,n}$ was defined by recursion. In what follows, we offer an *explicit* formulation in terms of seed number Φ_m. The reader may first prove by induction that difference equation

$$F_{m,n+2} = F_{m,n} + \left(\Phi_m - \frac{1}{\Phi_m} \right) F_{m,n+1}$$

accepts solution

$$F_{m,n} = A \left(\Phi_m^n - \left(\frac{-1}{\Phi_m} \right)^n \right),$$

which implies that $F_{m,0} = 0$. Setting the "boundary condition" $F_{m,1} = 1$, we obtain the explicit formulation

$$F_{m,n} = \frac{\Phi_m^n - (-1/\Phi_m)^n}{\Phi_m + 1/\Phi_m}. \tag{3.4a}$$

Generally, putting $F_{m,0} = 0$ and $F_{m,1} = k$ results in

$$F_{m,n} = k \frac{\Phi_m^n - (-1/\Phi_m)^n}{\Phi_m + 1/\Phi_m}. \tag{3.4b}$$

Using equations (3.2) and (3.3), expression (3.4a) becomes

$$F_{m,n} = \frac{1}{\sqrt{4 + m^2}} \left(\left(\frac{m + \sqrt{4 + m^2}}{2} \right)^n - \left(\frac{m - \sqrt{4 + m^2}}{2} \right)^n \right). \tag{3.5}$$

In the particular case of $m = 1$, we rediscover the expression originally discovered by de Moivre in 1718:

$$F_{1,n} = \frac{1}{\sqrt{5}} \left(\left(\frac{1 + \sqrt{5}}{2} \right)^n - \left(\frac{1 - \sqrt{5}}{2} \right)^n \right). \tag{3.6a}$$

For $m = 2$, we rediscover the following expression, originally discovered by the English mathematician John Pell (1610–1685):

$$F_{2,n} = \frac{1}{2\sqrt{2}} ((1 + \sqrt{2})^n - (1 - \sqrt{2})^n). \qquad (3.6b)$$

As a further exercise the reader may also prove the important statement

$$\Phi_m^n = F_{m,n-1} + F_{m,n}\Phi_m, \qquad (3.7)$$

from which it follows that

$$\sqrt[n]{F_{m,n-1} + F_{m,n}\sqrt[n]{F_{m,n-1} + F_{m,n}\sqrt[n]{F_{m,n-1} + F_{m,n}\sqrt[n]{\cdots}}}} \to \Phi_m.$$

$$(3.8a)$$

In particular,

$$\sqrt{1 + m\sqrt{1 + m\sqrt{1 + m\sqrt{\cdots}}}} \to \Phi_m. \qquad (3.8b)$$

It also follows that

$$\Phi_m = \frac{F_{m,n} + F_{m,n+1}\Phi_m}{F_{m,n-1} + F_{m,n}\Phi_m}, \quad \text{for any } n. \qquad (3.8c)$$

In chaper VI, we shall explore the case of $m = 1$. For $m = 2$, we get $\Phi_2 = 1 + \sqrt{2}$, and equation (3.7) yields

$$\Phi_2^n = F_{2,n-1} + F_{2,n}\Phi_2.$$

The successive non-negative powers of Φ_2 are therefore

$$1, \Phi_2, 1 + 2\Phi_2, 2 + 5\Phi_2, 5 + 12\Phi_2, 12 + 29\Phi_2, \ldots \qquad (3.9a)$$

The integral coefficients in the above sequence belong to the Fibonacci sequence of order $m = 2$, namely, $0, 1, 2, 5, 12, 29, \ldots$. The sequence may be extended to negative powers of Φ_2, yielding the sequence

$$\ldots, 5 - 2\Phi_2, -2 + \Phi_2, 1, \Phi_2, 1 + 2\Phi_2, 2 + 5\Phi_2, \ldots, \qquad (3.9b)$$

whose numerical values are

$$\ldots, 3 - 2\sqrt{2}, -1 + \sqrt{2}, 1, 1 + \sqrt{2}, 3 + 2\sqrt{2}, 7 + 5\sqrt{2}, \ldots \qquad (3.9c)$$

It will be observed that if T_i, T_{i+1}, T_{i+2} are any three consecutive terms of the sequence, we have

$$\Phi_2 T_i = T_{i+1} \quad \text{and} \quad T_i + 2T_{i+1} = T_{i+2}.$$

In turn, equation (3.8) yields

$$\Phi_2 = \frac{1 + 2\Phi_2}{\Phi_2} = \frac{2 + 5\Phi_2}{1 + 2\Phi_2} = \frac{5 + 12\Phi_2}{2 + 5\Phi_2} = \frac{12 + 20\Phi_2}{5 + 12\Phi_2} = \cdots$$

or

$$= \frac{3 - 2\sqrt{2}}{-7 + 5\sqrt{2}} = \frac{-1 + \sqrt{2}}{3 - 2\sqrt{2}} = \frac{1}{-1 + \sqrt{2}}$$

$$= 1 + \sqrt{2} = \frac{3 + 2\sqrt{2}}{1 + \sqrt{2}} = \frac{2 + 5\sqrt{2}}{3 + 2\sqrt{2}} = \cdots \quad (3.10)$$

These results may be generalized to any m, thus

$$\Phi_m T_i = T_{i+1} \quad \text{and} \quad T_i + m T_{i+1} = T_{i+2}.$$

The particular case of $m = 1$ will be addressed upon studying the golden number. Putting

$$\psi_{m,n} = \frac{F_{m,n+1}}{F_{m,n}}, \quad (3.11)$$

equations (3.4) yield, for large values of n,

$$\psi_{m,n} = \frac{F_{m,n+1}}{F_{m,n}} = \frac{\Phi_m^{n+1} - (-1/\Phi_m)^{n+1}}{\Phi_m^n - (-1/\Phi_m)^n} \approx \Phi_m$$

or

$$\psi_{m,n} = \frac{F_{m,n+1}}{F_{m,n}} \xrightarrow[n \ large]{} \Phi_m. \quad (3.12a)$$

Below, we shall establish that $F_{m,n} = N_{n-2} = D_{n-1}$, where N_n and D_n respectively represent the numerator and denominator of the nth convergent of continued fraction $[m, m, m, \ldots]$. Huygens's theorem which was presented in chapter 2, establishes that $N_{n-1}D_n - N_i D_{\tilde{n}-1} = (-1)^n$. Thus,

$$(F_{m,n+1})^2 - F_{m,n}F_{m,n+2} = (-1)^n, \quad \text{that is,} \quad (F_{m,n+1})^2 = F_{m,n}F_{m,n+2} + (-1)^n,$$

and we obtain

$$\psi_{m,n} - \frac{1}{\psi_{m,n}} = \frac{F_{m,n+1}}{F_{m,n}} - \frac{F_{m,n}}{F_{m,n+1}} = \frac{(F_{m,n+1})^2 - (F_{m,n})^2}{F_{m,n}F_{m,n+1}}$$

$$= \frac{F_{m,n}(F_{m,n+2} - F_{m,n}) + (-1)^n}{F_{m,n}F_{m,n+1}} = \frac{m(F_{m,n}F_{m,n+1}) + (-1)^n}{F_{m,n}F_{m,n+1}},$$

$$\psi_{m,n} - \frac{1}{\psi_{m,n}} = m + \frac{(-1)^n}{F_{m,n}F_{m,n+1}},$$

hence

$$\psi_{m,n} - \frac{1}{\psi_{m,n}} \xrightarrow[n\ large]{} m. \qquad (3.12b)$$

For example,

$$\psi_{1,6} = \frac{13}{8}, \qquad \frac{13}{8} - \frac{8}{13} = \frac{105}{104} = 1 + \frac{1}{104} \approx 1,$$

$$\psi_{2,5} = \frac{70}{29}, \qquad \frac{70}{29} - \frac{29}{70} = \frac{4{,}059}{2{,}030} = 2 - \frac{1}{2{,}030} \approx 2,$$

and we also have, from equations (3.4),

$$\frac{\Phi_m^{n+1}}{\Phi_m^2 + 1} = \frac{\Phi_m^{n+1}}{m\Phi_m + 2} \xrightarrow[n\ large]{} F_{m,n}. \qquad (3.12c)$$

For example,

$$\frac{\Phi_1^{16}}{\Phi_1 + 2} \approx \frac{2{,}206}{3.618034} \approx 609.7234 \approx 610 = F_{1,15},$$

$$\frac{\Phi_2^8}{2\Phi_2 + 2} \approx \frac{1{,}154}{6.828427} \cong 169 = F_{2,7}.$$

As an exercise, the reader may verify that the above result holds true no matter which value is assigned to $F_{m,1}$, other than 1 in definition

(3.1). Using formulation (3.4a) for $F_{m,n}$, the reader can also show that

$$F_{m,2n+1} = F_{m,n}^2 + F_{m,n+1}^2, \tag{3.13}$$

$$F_{m,2n+2} = \frac{F_{m,n+2}^2 - F_{m,n}^2}{m}. \tag{3.14}$$

Statements (3.13) and (3.14) can be used, together with initial conditions $F_{m,0} = 0, F_{m,1} = 1$, as an alternative recursive method for generating Fibonacci sequences.

ALTERNATIVE EXPLICIT FORMULATION

Here we offer yet another explicit formulation for $F_{m,i}$. Let $\binom{i}{j}$ denote the *binomial coefficient*

$$\frac{i!}{j!(i-j)!}. \tag{3.15}$$

A fundamental property of that function of integers i and j is

$$\binom{i+1}{j+1} = \binom{i}{j} + \binom{i}{j+1}, \quad i = 0, 1, 2, \ldots. \tag{3.16}$$

We now define polynomial $P_i(x)$ as

$$P_i(x) = \sum_j \binom{i-j}{j} x^j, \quad j = 0, 1, 2, \ldots, i; \tag{3.17a}$$

using property (3.16), it may be easily shown that

$$P_{i+1}(x) = xP_{i-1}(x) + P_i(x), \quad i = 0, 1, 2, \ldots. \tag{3.17b}$$

This in turn allows us to write, with $F_{m,0} = 0$,

$$F_{m,i+1} = m^i P_i\left(\frac{1}{m^2}\right). \tag{3.18a}$$

This equation may be proven by induction as follows:

56

TABLE 3.2
The Binomial Coefficients

	$j \longrightarrow$								$j \longrightarrow$			
	0	1	2	3	4	5	6		0	1	2	3
0	1							0	1			
1	1	1						1	1			
2	1	2	1					2	1	1		
3	1	3	3	1				3	1	2		
4	1	4	6	4	1			4	1	3	1	
5	1	5	10	10	5	1		5	1	4	3	
6	1	6	15	20	15	6	1	6	1	5	6	1
7	1	7	21	35	35	...		7	1	6	10	4

i (left table, vertical arrow down) i (right table, vertical arrow down)

$$\binom{i}{j} \qquad\qquad \binom{i-j}{j}$$

1. It is easily verified for $i = 0$ and $i = 1$.
2. From the definition of equation (3.1), and assuming the equation to be verified for $i = n$ and $i = n +1$, we can write

$$F_{m,n+2} = m^{n-1} P_{n-1}\left(\frac{1}{m^2}\right) + m^{n+1} P_n\left(\frac{1}{m^2}\right)$$

$$= m^{n+1}\left[\frac{1}{m^2} P_{n-1}\left(\frac{1}{m^2}\right) + P_n\left(\frac{1}{m^2}\right)\right],$$

$$F_{m,n+2} = m^{n+1} P_{n+1}\left(\frac{1}{m^2}\right). \qquad\qquad (3.18b)$$

Statement (3.18a) is thus verified for $i = n +2$ and is therefore true for all values of i. It can be explicited as follows:

$$F_{m,n+1} = m^n \sum_j \binom{n-j}{j}\left(\frac{1}{m^2}\right)^j, \quad i = 0, 1, 2, \dots \qquad (3.18c)$$

Examples. Using table 3.2 of coefficients $\binom{i-j}{j}$, we can calculate the following values of $F_{m,j}$:

$$F_{1,5} = 1 + 3 + 1 = 5, \qquad\qquad F_{3,3} = 3^2(1 + 1/9) = 10,$$

$$F_{2,3} = 2^2(1 + 1/4) = 5, \qquad\qquad F_{3,4} = 3^3(1 + 2/9) = 33,$$

$$F_{2,4} = 2^3(1 + 2/4) = 12, \qquad\qquad F_{3,5} = 3^4(1 + 3/9 + 1/81) = 109.$$

$$F_{2,5} = 2^4(1 + 3/4 + 1/16) = 29,$$

Note that the sum of the integers on any line i of the right-hand table is equal to $F_{1,i}$. The reader may also want to verify the following statements:

$$\left.\begin{aligned}
F_{m,2n} &= \sum_{j=0}^{n-1} \binom{n+j}{2j+1} m^{2j+1} \\
F_{m,2n+1} &= \sum_{j=0}^{n} \binom{n+j}{2j} m^{2j}
\end{aligned}\right\}, n,j = 0, 1, 2, \ldots \qquad (3.18d)$$

THE MONOGNOMONIC SIMPLE PERIODIC FRACTION

By definition, regular continued fractions have integral quotients. Nothing precludes, however, the use of simple continued fractions (with unit numerators) with nonintegral or even complex "quotients," as we shall see later. Although the term quotient carries an integral connotation within the framework of the division algorithm, we shall nonetheless continue using that term in the general case. The continued fractions at hand, although simple, can no longer be referred to as regular. We shall nonetheless take the liberty of writing bracketed expresions such as

$$\frac{\sqrt{5}+1}{2} = \left[1, 1, \ldots, 1, \frac{\sqrt{5}+1}{2}\right], \qquad (3.19a)$$

$$\sqrt{2} = \left[\frac{1}{\sqrt{2}}, \frac{1}{\sqrt{2}}, \frac{1}{\sqrt{2}}, \frac{1}{\sqrt{2}}, \ldots\right], \qquad (3.19b)$$

$$\frac{\sqrt{3}+1}{2} = \left[1, 2, 1, 2, \ldots, 1, 2, \frac{\sqrt{3}+1}{2}\right], \qquad (3.19c)$$

$$\frac{\sqrt{3}+1}{2} = [1, 2, 1, 2, \ldots, 1, \sqrt{3}+1]. \qquad (3.19d)$$

Notwithstanding the terminations, if any, the bracketed expressions are strictly periodic simple periodic continued fractions, SPF for short. The first two SPF's, of period 1, are said to be *monognomonic*. The latter two, of period 2, are said to be *dignomonic*. Generally, an SPF of period N shall be said to be N-gnomonic.

Here we shall be addressing only dignomonic and monognomonic SPFs, because of their relevance to the study of certain classes of iterative processes, including electrical ladder circuits and spirals. Comparing recursive formulae (2.12) of continued fractions:

$$N_i = N_{i-2} + q_i N_{i-1}, \qquad D_i = D_{i-2} + q_i D_{i-1}, \quad \text{for } i = 0, 1, 2, \ldots,$$

where

$$N_{-1} = D_{-2} = 1, \qquad N_{-2} = D_{-1} = 0 \qquad (3.20a)$$

with recursive formula

$$F_{m,i+2} = F_{m,i} + m F_{m,i+1}, \quad \text{where } F_{m,0} = 0, F_{m,1} = 1,$$
$$(3.20b)$$

and putting $q_0 = q_1 = q_2 \ldots = q_{i-1} = m$, it follows that

$$F_{m,i} = N_{i-2} = D_{i-1}, \qquad (3.20c)$$

and

$$\delta_i = \frac{N_{i-2} + q_i N_{i-1}}{D_{i-2} + q_i D_{i-1}} = \frac{F_{m,i} + q_i F_{m,i+1}}{F_{m,i-1} + q_i F_{m,i}}. \qquad (3.20d)$$

Consider now the SPF

$$\phi = [\underbrace{m, m, m, \quad \ldots \qquad \qquad m, m, \ldots, m}_{n}, \phi_\tau], \qquad (3.21)$$

where positive number ϕ_τ, rational or irrational, is referred to as the *termination*. Starting at the far left with increasing values of index i, its convergents are

$$\delta_0 = [m] = \frac{F_{m,0} + mF_{m,1}}{F_{m,-1} + mF_{m,0}} = \frac{F_{m,2}}{F_{m,1}} = m,$$

$$\delta_1 = [m, m] = \frac{F_{m,1} + mF_{m,2}}{F_{m,0} + mF_{m,1}} = \frac{F_{m,3}}{F_{m,2}},$$

$$\cdots$$

$$\delta_i = \underbrace{[m, m, m, \ldots, m]}_{i+1} = \frac{F_{m,i} + mF_{m,i+1}}{F_{m,i-1} + mF_{m,i}} = \frac{F_{m,i+2}}{F_{m,i+1}}$$

$$\cdots \quad . \tag{3.22}$$

Starting from the termination, which may thus be indifferently referred to as the *seed*, we can write, calling ρ_j the *jth retroconvergent*, or *retrovergent*,

$$\rho_0 = \phi_\tau,$$

$$\rho_1 = [m, \phi_\tau] = \frac{F_{m,1} + \phi_\tau F_{m,2}}{F_{m,0} + \phi_\tau F_{m,1}},$$

$$\rho_2 = [m, m, \phi_\tau] = \frac{F_{m,2} + \phi_\tau F_{m,1}}{F_{m,1} + \phi_\tau F_{m,2}},$$

$$\cdots$$

$$\rho_j = \left[\underbrace{m, m, m, \ldots, m}_{j}, \phi_\tau \right] = \frac{F_{m,j} + \phi_\tau F_{m,j+1}}{F_{m,j-1} + \phi_\tau F_{m,j}}$$

$$\cdots \quad . \tag{3.23}$$

and we get

$$\phi = \rho_n = \delta_n = \left[\underbrace{m, m, m, \ldots, m}_{n}, \phi_\tau \right] = \frac{F_{m,n} + \phi_\tau F_{m,n+1}}{F_{m,n-1} + \phi_\tau F_{m,n}}. \tag{3.24}$$

As an exercise, the reader may prove that if termination

$$\phi_\tau = \psi_{m,k} = \frac{F_{m,k+1}}{F_{m,k}}$$

we get

$$\left[\underbrace{m, m, m, \ldots, m}_{n}, \psi_{m,k} \right] = \psi_{m,k+n}.$$

The monognomonic SPF is *properly terminated (seeded)* when $\phi_\tau = \Phi_m$, as defined by equation (3.2). It follows that

$$\Phi_m = [m, \Phi_m] = [m, [m, \Phi_m]] = [m, m, \Phi_m] \qquad (3.25)$$

and for any value of n,

$$\Phi_m = \left[\underbrace{m, m, m, \ldots, m}_{n}, \Phi_m \right]. \qquad (3.26)$$

THE DIGNOMONIC SIMPLE PERIODIC FRACTION

Consider the two numbers

$$\phi_0 = \frac{\sqrt{5} + \sqrt{3}}{\sqrt{3}}, \qquad \phi_1 = \frac{\sqrt{3}}{\sqrt{5} - \sqrt{3}}.$$

They are found to be the solutions of the equations

$$\phi_0 = 2 + \frac{1}{\phi_1} = [2, \phi_1], \qquad \phi_1 = 3 + \frac{1}{\phi_0} = [3, \phi_0], \quad (3.27)$$

from which the following six dignomonic SPF's may be derived:

$$\phi_0 = [2, 3, 2, 3, \ldots], \qquad (3.28a)$$

$$\phi_1 = [3, 2, 3, 2, \ldots], \qquad (3.28b)$$

$$\phi_0 = [2, 3, 2, 3, \ldots, 2, 3, \phi_0], \qquad (3.28c)$$

$$\phi_0 = [2, 3, 2, 3, \ldots, 3, 2, \phi_1], \qquad (3.28d)$$

$$\phi_1 = [3, 2, 3, 2, \ldots, 2, 3, \phi_0], \qquad (3.28e)$$

$$\phi_1 = [3, 2, 3, 2, \ldots, 3, 2, \phi_1]. \qquad (3.28f)$$

SPFs (3.28a) and (3.28b) are nonterminating. Equations (3.28c) and (3.28f) suggest that starting from the last partial quotient, or termination, and working upstream, the retrovergent is equal to the termination following any even number of iterations. Equations (3.28d) and (3.28e), on the other hand, suggest that any odd number of iterations take the retrovergent from either termination value to the alternate value.

Equations (3.28) can be rewritten in general terms using special symbols which we refer to as *generic*. Generally, we shall let α represents the initial, α' the second, α the third, and so on, partial quotients. Similarly, ω represents the penultimate, and ω' the last partial quotients preceding arbitrary termination ϕ_τ, as in (3.29). Given partial quotients q_0, q_1, each of symbols α, α' may be assigned to either quotient, with the understanding that if α stands for q_0, then α' stands for q_1, and vice versa. *Independently of that assignment*, each of ω and ω' may also be assigned to either of q_0 and q_1 with a similar understanding, and we shall write

$$\phi = [\alpha, \alpha', \alpha, \ldots, \omega, \omega', \phi_\tau]. \tag{3.29}$$

Additionally, we shall define $\phi_{\alpha,\alpha'}$ and $\phi_{\alpha',\alpha}$ as follows:

$$\phi_{\alpha,\alpha'} = \alpha + \frac{1}{\phi_{\alpha',\alpha}}, \qquad \phi_{\alpha',\alpha} = \alpha' + \frac{1}{\phi_{\alpha,\alpha'}}. \tag{3.30a}$$

Because of the generic character of $\phi_{\alpha,\alpha'}$ and $\phi_{\alpha',\alpha}$, one of the above two statements is actually redundant. We may now write

$$\phi_{\alpha,\alpha'} = [\alpha, \phi_{\alpha',\alpha}] = [\alpha, \alpha', \phi_{\alpha,\alpha'}] \tag{3.30b}$$

$$= [\alpha, \alpha', \ldots, \omega, \omega', \phi_{\omega,\omega'}] \tag{3.30c}$$

$$= [\alpha, \alpha', \alpha, \alpha', \ldots]. \tag{3.30d}$$

It follows from (3.30a) that

$$\phi_{\alpha,\alpha'}^2 - \alpha\phi_{\alpha,\alpha'} - \frac{\alpha}{\alpha'} = 0. \tag{3.31}$$

Putting $\alpha\alpha' = \omega\omega' = m^2$ and solving for real positive values of $\phi_{\alpha,\alpha'}$, we get

$$\phi_{\alpha,\alpha'} = \alpha D_m = \frac{\alpha}{m}\Phi_m = \sqrt{\frac{\alpha}{\alpha'}}\,\Phi_m, \tag{3.32a}$$

$$\Phi_m^2 = \phi_{\alpha,\alpha'}\phi_{\alpha',\alpha}, \tag{3.32b}$$

$$\alpha\phi_{\alpha,\alpha'} = m\Phi_m = \Phi_m^2 - 1, \tag{3.32c}$$

where

$$\Phi_m = mD_m, \qquad D_m = \frac{1}{2}\left(1 + \sqrt{1 + \frac{4}{m^2}}\right). \tag{3.33}$$

Dignomonic SPF (3.29) is said to be *properly terminated* when $\phi_\tau = \phi_{\omega,\omega'}$, as in expressions (3.19a, c, d) (3.28c, d, e, f), and (3.30b, c).

ARBITRARILY TERMINATED SIMPLE PERIODIC FRACTIONS

Equation (3.34), which is a restatement of result (3.24), expresses retrovergent ρ_n for an arbitrarily terminated monognomonic SPF

$$\phi = \rho_n = \delta_n = \left[\underbrace{m, m, m, \ldots, m,}_{n}\; \phi_\tau\right] = \frac{F_{m,n} + \phi_\tau F_{m,n+1}}{F_{m,n-1} + \phi_\tau F_{m,n}} \tag{3.34}$$

For example,

$$\phi = [1, 1, 1, 1, 1, 2] = \frac{F_{1,5} + 2F_{1,6}}{F_{1,4} + 2F_{1,5}} = \frac{5 + 16}{3 + 10} = \frac{21}{13} \approx 1.6153846.$$

For large values of n, regardless of termination ϕ_τ, this becomes

$$\rho_n = \left[\underbrace{m, m, m, \ldots, m,}_{n}\; \phi_\tau\right] = \frac{1 + \phi_\tau\psi_{m,n}}{\phi_\tau + 1/\psi_{m,n-1}} \approx \Phi_m. \tag{3.35}$$

The above convergence is achieved fairly quickly when $m > 1$. We may now examine the following *arbitarily terminated dignomonic*

SPF, where ϕ_τ is an arbitrary positive number

$$\phi = [\alpha, \alpha', \alpha, \alpha', \ldots, \mu, \mu', \mu, \mu', \ldots, \omega, \omega', \phi_\tau], \quad (3.36)$$
$$\underset{\xleftarrow{\hspace{1cm}} \quad n \quad \xrightarrow{\hspace{1cm}}}{}$$

Case 1. n is even, that is, $\alpha = \omega$. We write

$$\phi = \frac{\alpha}{\sqrt{\alpha\alpha'}} \left[\underset{\xleftarrow{\hspace{1.5cm}} n \xrightarrow{\hspace{1.5cm}}}{\sqrt{\alpha\alpha'}, \sqrt{\alpha\alpha'}, \ldots, \sqrt{\alpha\alpha'}}, \frac{\sqrt{\alpha\alpha'}}{\alpha} \phi_\tau \right],$$

putting $\sqrt{\alpha\alpha'} = \sqrt{\omega\omega'} = m$ we get

$$\phi = \sqrt{\frac{\alpha}{\alpha'}} \left[\underset{\xleftarrow{\hspace{1cm}} n \xrightarrow{\hspace{1cm}}}{m, m, m, \ldots, m}, \sqrt{\frac{\omega'}{\omega}} \phi_\tau \right]. \quad (3.37)$$

Case 2. n is odd, that is, $\alpha' = \omega$. We write

$$\phi = \frac{\alpha}{\sqrt{\alpha\alpha'}} \left[\underset{\xleftarrow{\hspace{1.5cm}} n \xrightarrow{\hspace{1.5cm}}}{\sqrt{\alpha\alpha'}, \sqrt{\alpha\alpha'}, \ldots, \sqrt{\alpha\alpha'}}, \frac{\alpha}{\sqrt{\alpha\alpha'}} \phi_\tau \right],$$

$$\phi = \sqrt{\frac{\alpha}{\alpha'}} \left[\underset{\xleftarrow{\hspace{1cm}} n \xrightarrow{\hspace{1cm}}}{m, m, m, \ldots, m}, \sqrt{\frac{\omega'}{\omega}} \phi_\tau \right]. \quad (3.38)$$

Hence, the following important result holds for any value of n:

$$\phi = \underset{\xleftarrow{\hspace{1cm}} n \xrightarrow{\hspace{1cm}}}{[\alpha, \alpha', \alpha, \ldots, \omega, \omega', \phi_\tau]} = \sqrt{\frac{\alpha}{\alpha'}} \left[\underset{\xleftarrow{\hspace{1cm}} n \xrightarrow{\hspace{1cm}}}{m, m, m, \ldots, m}, \sqrt{\frac{\omega'}{\omega}} \phi_\tau \right]$$

$$= \sqrt{\frac{\alpha}{\alpha'}} \frac{F_{m,n} + F_{m,n+1} \sqrt{\dfrac{\omega'}{\omega}} \phi_\tau}{F_{m,n-1} + F_{m,n} \sqrt{\dfrac{\omega'}{\omega}} \phi_\tau}. \quad (3.39)$$

We can now write, starting from the left-hand end of (3.36),

$$\delta_i = \underset{\xleftarrow{\hspace{1cm}} i+1 \xrightarrow{\hspace{1cm}}}{[\alpha, \alpha', \alpha, \ldots, \mu, \mu']} = \sqrt{\frac{\alpha}{\alpha'}} \underset{\xleftarrow{\hspace{1cm}} i+1 \xrightarrow{\hspace{1cm}}}{[m, m, m, \ldots, m]} = \sqrt{\frac{\alpha}{\alpha'}} \psi_{m,i+1}. \quad (3.40)$$

Example. To calculate the value of

$$\phi = \left[1, \frac{1}{2}, 1, \frac{1}{2}\right].$$

Turning to table 3.1, we have, with $m = 1/\sqrt{2}$,

$$\phi = \sqrt{\frac{1}{1/2}} \frac{F_{(1/\sqrt{2}),5}}{F_{(1/\sqrt{2}),4}} = \sqrt{2} \times \frac{11}{4} \times \frac{2\sqrt{2}}{5} = \frac{11}{5},$$

and we verify that, starting from the right-hand end of (3.36),

$$1 + \frac{1}{\dfrac{1}{2} + \dfrac{1}{1+2}} = \frac{11}{5}.$$

We also get

$$\rho_j = [\underbrace{\mu, \mu', \ldots, \omega, \omega', \phi_\tau}_{j}] = \sqrt{\frac{\mu}{\mu'}} \left[\underbrace{m, m, m, \ldots, m,}_{n} \sqrt{\frac{\omega'}{\omega}} \phi_\tau\right]$$

$$= \sqrt{\frac{\mu}{\mu'}} \frac{F_{m,j} + F_{m,j+1} \sqrt{\dfrac{\omega'}{\omega}} \phi_\tau}{F_{m,j-1} + F_{m,j} \sqrt{\dfrac{\omega'}{\omega}} \phi_\tau}. \tag{3.41}$$

Example. To calculate the value of $\phi = \left[1, \dfrac{1}{2}, 1, \dfrac{1}{2}, \sqrt{2}\right]$:

$$\phi = \sqrt{2} \frac{5/(2\sqrt{2}) + 11/4}{3/2 + 5/(2\sqrt{2})} = \frac{5\sqrt{2} + 11}{3\sqrt{2} + 5}.$$

When n is large, regardless of termination ϕ_T,

$$\phi = \left[\underbrace{(\alpha, \alpha', \alpha, \alpha', \ldots, \omega, \omega')}_{n}, \phi_T \right] \approx \sqrt{\frac{\alpha}{\alpha'}} \, \Phi_m \approx \phi_{\alpha,\alpha'}. \quad (3.42)$$

When we come to whorled figures, a striking geometric illustration of that statement will be given.

m IS VERY SMALL: FROM FIBONACCI TO HYPERBOLIC AND TRIGONOMETRIC FUNCTIONS

The hypothesis that m is very small, though somewhat academic at first sight, will prove surprisingly useful upon discussing certain physical phenomena, such as the electrical behavior of transmission lines. Let us first turn to Φ_m. Recalling equation (3.3a),

$$\Phi_m = \frac{\sqrt{4 + m^2} + m}{2}, \qquad \frac{1}{\Phi_m} = \frac{\sqrt{4 + m^2} - m}{2}.$$

When m is very small, we may neglect m^2 compared to 4, and we obtain

$$\Phi_m \approx 1 + \frac{m}{2}, \qquad \frac{1}{\Phi_m} \approx 1 - \frac{m}{2}. \quad (3.43a)$$

Equation (3.4a) can now be rewritten for even values of n

$$F_{m,n} = \frac{\Phi_m^n - (-1/\Phi_m)^n}{\Phi_m + 1/\Phi_m} = \frac{\Phi_m^n - 1/\Phi_m^n}{\Phi_m + 1/\Phi_m} \approx \frac{\left(1 + \dfrac{m}{2}\right)^n - \left(1 - \dfrac{m}{2}\right)^n}{1 + \dfrac{m}{2} + 1 - \dfrac{m}{2}}.$$

The smallness of $m/2$ allows us to put

$$\left(1 + \frac{m}{2}\right)^n = e^{\frac{m}{2} \times n} = e^{\frac{mn}{2}} \quad \text{and} \quad \left(1 - \frac{m}{2}\right)^n = e^{-\frac{mn}{2}}; \quad (3.43b)$$

hence, for even values of n,

$$F_{m,n} \approx \frac{e^{\frac{mn}{2}} - e^{-\frac{mn}{2}}}{2} = \sinh \frac{mn}{2}. \tag{3.43c}$$

Following the same reasoning, we get for odd values of n

$$F_{m,n} \approx \frac{e^{\frac{mn}{2}} + e^{-\frac{mn}{2}}}{2} = \cosh \frac{mn}{2}. \tag{3.43d}$$

It is well known that $\sinh jx = j \sin x$ and $\cosh jx = \cos x$. If we replace m by jm in equations (3.43c) and (3.43d), we obtain

$$F_{jm,n} \approx j \sin \frac{mn}{2} \quad \text{for even values of } n, \tag{3.43e}$$

$$F_{jm,n} \approx \cos \frac{mn}{2} \quad \text{for odd values of } n. \tag{3.43f}$$

APPENDIX: THE POLYGNOMONIC SPF

As stated earlier, we shall dwell in what follows only upon monognomonic and dignomonic SPFs. Nonetheless, we observe that

$$\phi_{\alpha,\alpha',\alpha'',\alpha''',\ldots,\alpha^{(n)}} = [\alpha, \alpha', \alpha'', \alpha''', \ldots, \omega, \phi_{\omega',\omega'',\omega''',\ldots,\omega^{(n)},\omega}], \tag{3.44}$$

$$\phi_{\alpha,\alpha',\alpha'',\alpha''',\ldots,\alpha^{(n)}} = \alpha + \frac{1}{\phi_{\alpha',\alpha'',\alpha''',\ldots,\alpha^{(n)},\alpha}}. \tag{3.45}$$

As an exercise, the interested reader may consider the *tetragnomonic* SPF, that is, one which corresponds to a continued fraction of period 4, where α, α', α'', α''' are *generic* symbols such that if α represents any given partial quotient, then α', α'', α''' represent the following three quotients in the periodic sequence, in their proper order. The same definition applies to ω, ω', ω'', ω''', irrespective of the assignment of values to α, α', α'', α'''. We get

$$\phi_{\alpha,\alpha',\alpha'',\alpha'''} = \alpha + \frac{1}{\phi_{\alpha',\alpha'',\alpha''',\alpha}},$$

$$\phi_{\alpha,\alpha',\alpha'',\alpha'''} = [\alpha, \alpha', \alpha'', \alpha''', \ldots, \omega, \phi_{\omega',\omega'',\omega''',\omega}],$$

and the reader may verify the following statements:

$$\phi_{1,2,1,6} = [1, 2, 1, 6, \ldots, 6, 1, 2, 1, 6, \phi_{1,2,1,6}]$$
$$= [1, 2, 1, 6, \ldots, 1, 2, 1, \phi_{6,1,2,1}]$$
$$= [1, 2, 1, 6, \ldots, 1, 2, \phi_{1,6,1,2}]$$
$$= [1, 2, 1, 6, \ldots, 1, \phi_{2,1,6,1}].$$

Similarly,

$$\phi_{2,1,6,1} = [2, 1, 6, \ldots, 6, 1, 2, 1, 6, \phi_{1,2,1,6}]$$
$$= [2, 1, 6, \ldots, 1, 2, 1, \phi_{6,1,2,1}]$$
$$= [2, 1, 6, \ldots, 1, 2, \phi_{1,6,1,2}]$$
$$= [2, 1, 6, \ldots, 1, \phi_{2,1,6,1}].$$

$$\cdots,$$

and in particular,

$$\phi_{1,2,1,6} = [1, \phi_{2,1,6,1}] = \frac{\sqrt{14} + 3}{5},$$

$$\phi_{2,1,6,1} = [2, \phi_{1,6,1,2}] = \frac{\sqrt{14} + 2}{2},$$

$$\phi_{1,6,1,2} = [1, \phi_{6,1,2,1}] = \frac{\sqrt{14} + 2}{5},$$

$$\phi_{6,1,2,1} = [6, \phi_{1,2,1,6}] = \sqrt{14} + 3.$$

The reader may also verify that the above statements are consistent with the spectrum $\sqrt{14} = [3, \overline{1, 2, 1, 6}]$, from which it follows that $1/(\sqrt{14} - 3) = (\sqrt{14} + 3)/5 = [\overline{1, 2, 1, 6}] = \phi_{1,2,1,6}$.

SUMMARY OF FORMULAE

Definitions and Recursive Formulations of $F_{m,n}$

$$F_{m,n+2} = F_{m,n} + mF_{m,n+1}, \quad \text{with } F_{m,0} = 0, F_{m,1} = 1,$$

$$m = \Phi_m - \frac{1}{\Phi_m},$$

$$\psi_{m,n} = \frac{F_{m,n+1}}{F_{m,n}},$$

$$D_m = \frac{1}{2}\left(1 + \sqrt{1 + \frac{4}{m^2}}\right).$$

Explicit Formulations of $F_{m,n}$

$$F_{m,n} = \frac{\Phi_m^n - (-1/\Phi_m)^n}{\Phi_m + 1/\Phi_m},$$

$$F_{m,n} = \frac{1}{\sqrt{4 + m^2}}\left(\left(\frac{m + \sqrt{4 + m^2}}{2}\right)^n - \left(\frac{m - \sqrt{4 + m^2}}{2}\right)^n\right),$$

$$F_{1,n} = \frac{1}{\sqrt{5}}\left(\left(\frac{1 + \sqrt{5}}{2}\right)^n - \left(\frac{1 - \sqrt{5}}{2}\right)^n\right),$$

$$F_{m,n+1} = m^n \sum_j \binom{n-j}{j}\left(\frac{1}{m^2}\right)^j,$$

$$F_{m,2n} = \sum_{j=0}^{n-1} \binom{n+j}{2j+1} m^{2j+1},$$

$$F_{m,2n+1} = \sum_{j=0}^{n-1} \binom{n+j}{2j} m^{2j}, \quad j = 0, 1, 2, \ldots.$$

Properties of Φ_m

$$\Phi_m^2 - m\Phi_m - 1 = 0,$$

$$\Phi_m = mD_m,$$

$$\Phi_m = \frac{\sqrt{4+m^2}+m}{2}, \quad \frac{1}{\Phi_m} = \frac{\sqrt{4+m^2}-m}{2}, \quad \Phi_m + \frac{1}{\Phi_m} = \sqrt{4+m^2},$$

$$\Phi_m^2 = \frac{D_m}{D_m - 1}, \quad D_m = \frac{\Phi_m^2}{\Phi_m^2 - 1},$$

$$\Phi_m^n = F_{m,n-1} + F_{m,n}\Phi_m,$$

$$\Phi_m = \sqrt[n]{F_{m,n-1} + F_{m,n}\sqrt[n]{F_{m,n-1} + F_{m,n}\sqrt[n]{F_{m,n-1} + F_{m,n}\sqrt[n]{\ldots}}}},$$

$$\Phi_m = \sqrt{1 + m\sqrt{1 + m\sqrt{1 + m\sqrt{\ldots}}}},$$

$$\Phi_m = \frac{F_{m,n} + F_{m,n+1}\Phi_m}{F_{m,n-1} + F_{m,n}\Phi_m} \text{ for any } n,$$

$$\psi_{m,n} = \frac{F_{m,n+1}}{F_{m,n}} \xrightarrow[n \, large]{} \Phi_m,$$

$$\psi_{m,n} - \frac{1}{\psi_{m,n}} = m + \frac{(-1)^n}{F_{m,n}F_{m,n+1}},$$

$$\psi_{m,n} - \frac{1}{\psi_{m,n}} \xrightarrow[n \, large]{} m,$$

$$\frac{\Phi_m^{n+1}}{\Phi_{m+1}^2} = \frac{\Phi_m^{n+1}}{m\Phi_m + 2} \xrightarrow[n \, large]{} F_{m,n}.$$

The Monognomonic SPF

GENERAL FORM

$$\phi = \underset{|\leftarrow \qquad\quad n \qquad\quad \rightarrow|}{[m, m, m, \ldots, m, m, \ldots, m, \phi_\tau],}$$

$$\delta_i = \underset{\leftarrow \quad i+1 \quad \rightarrow}{[m, m, m, \ldots, m]} = \frac{F_{m,i} + mF_{m,i+1}}{F_{m,i-1} + mF_{m,i}} = \frac{F_{m,i+2}}{F_{m,i+1}} = \psi_{m,i+1},$$

$$\rho_j = \underset{\leftarrow \quad j \quad \rightarrow}{[m, m, m, \ldots, m, \phi_\tau]} = \frac{F_{m,j} + \Phi_\tau F_{m,j+1}}{F_{m,j-1} + \Phi_\tau F_{m,j}} = \frac{\phi_\tau \psi_{m,j} + 1}{\phi_\tau + 1/\psi_{m,j-1}}.$$

PROPERLY TERMINATED FORM: $\phi_\tau = \Phi_m$

$$\Phi_m = [\underleftrightarrow{m, m, m, \ldots, m}_{n}, \Phi_m] \text{ for all } n.$$

IMPROPERLY TERMINATED FORM: $\phi_\tau \neq \Phi_m$

$$\phi = \delta_n = \rho_n = [\underset{\underset{n}{\overset{\rightarrow}{\leftarrow}}}{m, m, m, \ldots, m}, \phi_\tau] = \frac{F_{m,n} + \Phi_\tau F_{m,n+1}}{F_{m,n-1} + \phi_\tau F_{m,n}} = \frac{\phi_\tau \psi_{m,n} + 1}{\phi_\tau + 1/\psi_{m,n-1}},$$

$$[\underset{\underset{n}{\overset{\rightarrow}{\leftarrow}}}{m, m, m, \ldots, m}, \phi_\tau] = \xrightarrow[n \, large]{} \Phi_m \text{ regardless of } \phi_\tau,$$

$$[\underset{\underset{n}{\overset{\rightarrow}{\leftarrow}}}{m, m, m, \ldots, m,}] = \psi_{m,n}$$

$$[\underset{\underset{n}{\overset{\rightarrow}{\leftarrow}}}{m, m, m, \ldots, m}, \psi_{m,k}] = \psi_{m,k+1}.$$

INFINITE FORM

$$[m, m, m, \ldots] = \Phi_m$$

The Dignomonic SPF

GENERAL FORM

$$\phi = [\alpha, \alpha', \alpha, \alpha', \ldots, \mu, \mu', \mu, \mu', \ldots, \omega, \omega', \phi_\tau],$$
$$| \underset{\underset{n}{\overset{\rightarrow}{\leftarrow}}}{} |$$

$$\delta_i = [\underset{\underset{i+1}{\overset{\rightarrow}{\leftarrow}}}{\alpha, \alpha', \alpha, \ldots, \mu, \mu'}] = \sqrt{\frac{\alpha}{\alpha'}} [\underset{\underset{i+1}{\overset{\rightarrow}{\leftarrow}}}{m, m, m, \ldots, m}] = \sqrt{\frac{\alpha}{\alpha'}} \, \psi_{m,i+1},$$

$$\rho_j = [\underset{\underset{j}{\overset{\rightarrow}{\leftarrow}}}{\mu, \mu', \ldots, \omega, \omega', \phi_\tau}] = \sqrt{\frac{\mu}{\mu'}} \left[\underset{\underset{j}{\overset{\rightarrow}{\leftarrow}}}{m, m, m, \ldots, m}, \sqrt{\frac{\omega'}{\omega}} \phi_\tau \right]$$

$$= \sqrt{\frac{\mu}{\mu'}} \frac{F_{m,j} + F_{m,j+1} \sqrt{\frac{\omega'}{\omega}} \phi_\tau}{F_{m,j-1} + F_{m,j} \sqrt{\frac{\omega'}{\omega}} \phi_\tau}.$$

71

PROPERLY TERMINATED FORM: $\phi_\tau = \phi_{\omega,\omega'}$

$$\alpha = \phi_{\alpha,\alpha'} - \frac{1}{\phi_{\alpha',\alpha}}, \qquad \alpha' = \phi_{\alpha',\alpha} - \frac{1}{\phi_{\alpha,\alpha'}}, \qquad \alpha\alpha' = \omega\omega' = m^2,$$

$$\phi_{\alpha,\alpha'} = [\underset{\underset{\leftarrow}{}}{\alpha, \alpha', \alpha, \dots} \underset{\underset{n}{}}{,} \underset{\underset{\rightarrow}{}}{\omega, \omega', \phi_{\omega,\omega'}}], \quad \text{for all } n,$$

$$\phi_{\alpha,\alpha'}^2 - \alpha\phi_{\alpha,\alpha'} - \frac{\alpha}{\alpha'} = 0,$$

$$\phi_{\alpha,\alpha'} = \alpha D_m = \frac{\alpha}{m}\Phi_m = \sqrt{\frac{\alpha}{\alpha'}}\,\Phi_m,$$

$$\Phi_m^2 = \phi_{\alpha,\alpha'}\,\phi_{\alpha',\alpha}, \quad \alpha\phi_{\alpha,\alpha'} = m\Phi_m = \Phi_m^2 - 1.$$

IMPROPERLY TERMINATED FORM: $\phi_\tau \neq \phi_{\omega,\omega'}$

$$\phi = \delta_n = \rho_n = [\underset{\leftarrow}{\alpha, \alpha',} \underset{n}{\dots,} \underset{\rightarrow}{\omega, \omega',} \phi_\tau] = \sqrt{\frac{\alpha}{\alpha'}}\left[\underset{\leftarrow}{m, m, m,} \underset{n}{\dots,} \underset{\rightarrow}{m,} \sqrt{\frac{\omega'}{\omega}}\,\phi_\tau\right]$$

$$= \sqrt{\frac{\alpha}{\alpha'}}\,\frac{F_{m,n} + F_{m,n+1}\sqrt{\dfrac{\omega'}{\omega}}\,\phi_\tau}{F_{m,n-1} + F_{m,n}\sqrt{\dfrac{\omega'}{\omega}}\,\phi_\tau},$$

$$[\underset{\leftarrow}{\alpha, \alpha', \alpha,} \underset{n}{\dots,} \underset{\rightarrow}{\omega, \omega',} \phi_\tau] = \xrightarrow[n\text{ large}]{} \sqrt{\frac{\alpha}{\alpha'}}\,\Phi_m = \phi_{\alpha,\alpha'}, \quad \text{regardless of } \phi_\tau,$$

$$[\underset{\leftarrow}{\alpha, \alpha',} \underset{n}{\dots,} \underset{\rightarrow}{\omega, \omega'}] = \sqrt{\frac{\alpha}{\alpha'}}\,[\underset{\leftarrow}{m, m, m,} \underset{n}{\dots,} \underset{\rightarrow}{m}] = \sqrt{\frac{\alpha}{\alpha'}}\,\psi_{m,n}.$$

INFINITE FORM

$$[\alpha, \alpha', \alpha, \alpha', \dots] = \phi_{\alpha,\alpha'}$$

When m Is Very Small

$$F_{m,n} \approx \sinh\frac{mn}{2} \text{ when } n \text{ is even, and } F_{m,n} \approx \cosh\frac{mn}{2} \text{ when } n \text{ is odd.}$$

When m Is Very Small and Imaginary

$$F_{jm,n} \approx j\sin\frac{mn}{2} \text{ when } n \text{ is even, and } F_{jm,n} \approx \cos\frac{mn}{2} \text{ when } n \text{ is odd.}$$

The Properly Terminated Polygnomonic SPF

$$\phi_{\alpha,\alpha',\alpha'',\alpha''',\ldots,\alpha^{(n)}} = [\alpha, \alpha', \alpha'', \alpha''', \ldots, \alpha^{(n)}, \alpha, \alpha', \alpha'', \alpha''', \ldots, \alpha^{(n)},$$
$$\alpha, \alpha', \alpha'', \ldots, \omega, \phi_{\omega',\omega'',\omega''',\ldots,\omega^{(n)},\omega}],$$

$$\phi_{\alpha,\alpha',\alpha'',\alpha''',\ldots,\alpha^{(n)}} = \alpha + \cfrac{1}{\phi_{\alpha',\alpha'',\alpha''',\ldots,\alpha^{(n)},\alpha}}.$$

Ladders: From Fibonacci to
Wave Propagation

This chapter constitutes an immediate application of continued fractions, and an interesting illustration of *iterative* processes involving *feedback*, ranging from electrical ladder construction to propagation along transmission lines.

A mechanical metaphor is also offered that uses pulleys and strings. It might constitute an interesting project for the amateur hobbyist or a college science fair contestant.

THE TRANSDUCER LADDER

Figure 4.1a represents a device akin to what engineers call an operational amplifier or OpAmp. It possesses two asymmetrical *input ports* and one *output port* and is characterized by a *transfer function* or *gain G*, which may be a real or imaginary number. We shall call that device a *transducer*. If "signals" x and y are applied to the inputs as shown, the resulting output signal z is given by $z = G(x - y)$. We are not concerned with the exact physical nature of the signals that are applied to the inputs and measured at the output. These may be voltages, currents, mechanical displacements, and so on. The transducer's role is to convert one kind of physical magnitude into a magnitude of another or of the same kind.

With the help of the transducer just described, the *recursive* structure of figure 4.1b may be constructed, whose successive transfer functions are μ_0, τ_0, μ_1, τ_2, ... With the various input and output ports labeled as shown in the figure, we can write

$$i_0 = (v_0 - v_1)\mu_1, \qquad v_1 = (i_0 - i_1)\tau_0,$$
$$i_1 = (v_1 - v_2)\mu_1, \qquad v_2 = (i_1 - i_2)\tau_1,$$
$$i_2 = (v_2 - 0)\mu_2. \qquad\qquad\qquad\qquad (4.1)$$

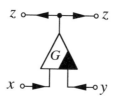

Fig. 4.1a. The transducer.

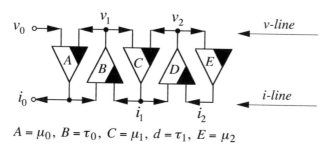

$A = \mu_0,\ B = \tau_0,\ C = \mu_1,\ d = \tau_1,\ E = \mu_2$

Fig. 4.1b. The transducer ladder.

These equations can be rewritten, putting $r_i = 1/\mu_1$ and $\rho_i = 1/\tau_i$,

$$\frac{v_0}{i_0} = r_0 + \frac{v_1}{i_0} \qquad \frac{i_0}{v_1} = \rho_0 + \frac{i_1}{v_1},$$

$$\frac{v_1}{i_1} = r_1 + \frac{v_2}{i_1} \qquad \frac{i_1}{v_2} = \rho_1 + \frac{i_2}{v_2},$$

$$\frac{v_2}{i_2} = r_2, \tag{4.2}$$

and we get

$$\frac{v_0}{i_0} = [r_0, \rho_0, r_1, \rho_1, r_2]. \tag{4.3}$$

Upon applying input v_0 to the top left input port of the network, output i_0 appears at the bottom left output port, and the ratio v_0/i_0 is given by equation (4.3).

Figure 4.1b provides some insight into the notion of feedback: signals v and i are fed back upstream into the precedings stages. That particular ladder feedback arrangement is nicely captured by continued fractions. One interesting property of the transducer diagram is that it highlights the existence of what might be called a *v-line* and an *i-line*, the interchanges between which are effected by the transducers.

THE ELECTRICAL LADDER

Without going into too much detail, we shall indicate simply that electrical behavior is defined in terms of two magnitudes, namely, *potential difference* and *current*. Each is uniquely and fully characterized by a value and a direction and may be represented by an arrow bearing a symbol denoting its value. An electrical circuit consists of interconnected elements, or components, that, for the purposes of the present study, are passive (they may not contain an internal source of energy) and bipolar (they are confined between two and only two extremities). These components are the resistor, the inductor, and the capacitor. A component's extremity, upon being electrically connected to that of another component, constitutes a *node*. The behavior of any particular component is defined in terms of the potential difference *across* the element, that is, between its extremities, and the current flowing *through* it. That behavior is governed by specific laws that depend upon the nature of the component. *Potential difference*, which is measured between any two nodes, is loosely referred to as *voltage difference* because its measurement unit is the *volt*. Electrical *current*, which *flows* through a circuit element, is measured in *amperes.*

Electrical circuits containing complex interconnections of elements obey two simple rules known as *Kirchhoff's rules.* Figure 4.2a shows current node N, together with currents i_0, i_1, i_2, i_3. Kirchhoff's first rule stipulates that the algebraic sum of the currents at any given node is zero. Thus, $i_0 + i_2 - i_1 - i_3 = 0$. Figure 4.2b shows an electrical circuit consisting of interconnected components. Potential differences

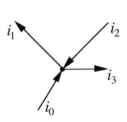

Fig. 4.2a. A current node.

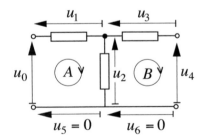

Fig. 4.2b. Voltage difference loops.

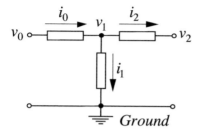

Fig. 4.2c. Voltages and currents.

$u_0, u_1, \ldots, u_6$ are represented by arrows standing on a base. Kirchoff's second rule stipulates that the algebraic sum of potential differences around a loop is equal to zero. Loop A yields $u_0 - u_1 - u_2 = 0$; in other words, $u_1 = u_0 - u_2$. Loop B yields $u_3 = u_2 - u_4$. Note the third redundant loop, which encompasses A and B, for which $u_1 + u_3 = u_0 - u_4$. This last statement can obviously be derived from the first two. Also note that the bottom two "components" are actually simple conductors, hence the statement $u_5 = u_6 = 0$. Generally, electrical engineers speak of *voltages* rather than voltage differences, where it is understood that a node's *voltage* is the *voltage difference* between that node and a common datum referred to as ground, whose voltage is defined to be zero. The figures we shall be dealing with are therefore akin to figure 4.2c, where $u_0 = v_0$, $u_1 = v_0 - v_1$, $u_2 = v_2$, $u_3 = v_1 - v_2$, $u_4 = v_4$, and $i_0 = i_1 + i_2$.

Resistance Ladders

Figure 4.2d shows the resistor, an electrical component that constitutes the cornerstone of all electronic circuits. Its simplest physical embodiment consists of a length of thin wire. Other forms consist of carbon or other resistive compounds. The most elaborate forms are found in integrated circuits, where they consist of regions of silicon substrate subjected to appropriate treatments. The resistor is characterized by the value of its *resistance* r, which is measured in *ohms*, whose symbol is Ω. The potential difference *across* the resistor, namely, $u =$

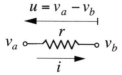

Fig. 4.2d. The resistor.

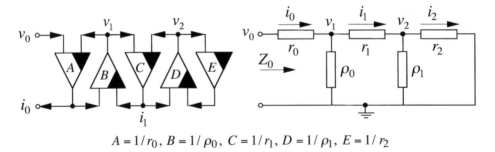

$$A = 1/r_0, \ B = 1/\rho_0, \ C = 1/r_1, \ D = 1/\rho_1, \ E = 1/r_2$$

Fig. 4.3a. Analogy between the transducer and electrical ladders.

$(v_a - v_b)$, drives current i inside the resistor, making it flow from the higher potential at a to the lower potential at b. That potential difference is related to current i by the most elementary law known to students of electricity, namely, *Ohm's law*, which stipulates that

$$i = \frac{u}{r} = \frac{v_a - v_b}{r}.$$

In addition to being passive, the resistor is said to be *linear*, by virtue of the linear relationship between u and i. The inverse of a resistance is a *conductance*. Electrical engineers, who are incredibly creative at times, have decided to call the measurement unit of conductances the *mho* (guess where it came from), whose symbol is an Ω turned upside down. Figure 4.3a, in which the little boxes represent resistors, shows an extremely common electrical circuit, known to electrical engineers as the *resistance ladder network*. The series (horizontal) arms are shown with the corresponding resistance values r_0, r_1, r_2, *measured in ohms*, whereas the parallel (vertical) arms are shown with the corresponding conductance values ρ_0, ρ_1, measured in mhos. All voltage differences shown in the diagram are measured with respect to the ground. Input voltage v_0 is applied as shown, causing current i_0 to flow in series arm r_0, as well as currents i_1, i_2 in series arms r_1, r_2. The reader will have no difficulty establishing the equations governing the behavior of that circuit, which are exactly those of the network of figure 4.2, namely, equations (4.2) and (4.3). The ratio $Z_0 = v_0/i_0 = [r_0, \rho_0, r_1, \rho_1, r_2]$ is known as the network's *input impedance* and is measured in ohms. In the numerical example of figure 4.3b,

$$r_0 = 4, \quad r_1 = 3, \quad r_2 = 2 \text{ ohms}, \quad \text{and } \rho_0 = 1, \quad \rho_1 = 2 \text{ mhos},$$

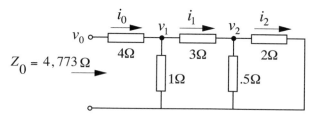

Fig. 4.3b. A resistance ladder.

and the network's input impedance (a resistance in the present case) is $[4, 1, 3, 2, 2] = 105/22 \approx 4.773\ \Omega$. Figure 4.3b suggests that Z_0 is the network's impedance as it would be "seen" from the input end by some hypothetical impedance-measuring device.

Iterative Ladders

We now turn our attention to figure 4.4, in which all the series resistances are equal to r and all the parallel conductances are equal to ρ. The ladder is said to be *iterative*. In the case of figure 4.4a, Z_t represents the network's terminating impedance. In the case of figure 4.4b, Y_t represents the terminating admittance. We get

$$Z_0 = \frac{v_0}{i_0} = [r, \rho, r, \rho, \ldots, r, \rho, Z_t], \quad \text{for figure 4.4a,}$$

and

$$Z_0 = \frac{v_0}{i_0} = [r, \rho, r, \rho, \ldots, r, Y_t], \quad \text{for figure 4.4b.}$$

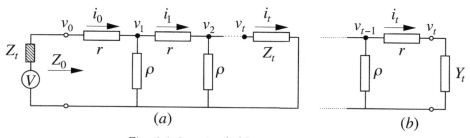

Fig. 4.4. Iterative ladder terminations.

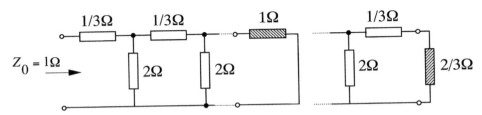

Fig. 4.5. An iterative resistance ladder and its proper terminations.

Putting

$$D_m = \frac{1}{2}\left(1 + \sqrt{1 + \frac{4}{r\rho}}\right), \quad \text{where } m = r\rho, \qquad (4.4)$$

the network is properly terminated when $Z_t = rD_m$, or when $Y_t = \rho D_m$, in both of which cases we get $Z_0 = Z_t$. If the power supply's "internal impedance" Z_t is also equal to Z_0, the source is said to be *matched* to the ladder.

Examples. In figure 4.5, $r = 1/3 \; \Omega$ and $\rho = 1/2$ mho (the reciprocal of 2Ω). Putting $m = r\rho$, we get

$$D_m = \frac{1}{2}\left(1 + \sqrt{1 + \frac{4}{r\rho}}\right) = \frac{1}{2}(1 + \sqrt{1 + 24}) = 3.$$

For the network to be properly terminated, we must make $Z_t = rD_m = 1\Omega$ as in figure 4.5a, or $Y_t = \rho D_m = 3/2$ mho (the reciprocal of a resistance of $2/3 \; \Omega$), as in figure 4.5b. In either case, input impedance $Z_0 = 1\Omega$. In figure 4.6, $r = 2\Omega$ and $\rho = 2$ mho. Thus, $D_m = (\sqrt{2} + 1)/2$. The proper termination shown in figure 4.6a is $2D_m = (\sqrt{2} + 1)\Omega$, and that shown in figure 4.6b is $Y_t = 2D_m = (\sqrt{2} + 1)$ mho, the reciprocal of $(\sqrt{2} - 1)\Omega$.

If the network comprises a large number of arms and is terminated by some arbitrary impedance, its input impedance converges to Z_0 as more arms are added, regardless of the actual terminating impedance. The input impedance of an "infinitely long" periodic ladder network, though by definition unterminated, tends to $Z_0 = rD_m$.

So far, we have examined the behavior of properly terminated periodic ladder networks from an input impedance point of view. We

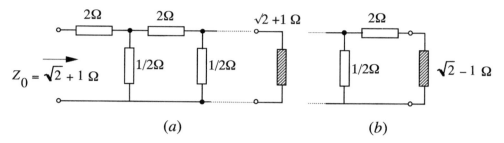

(a)   (b)

Fig. 4.6. Another properly terminated iterative resistance ladder.

now examine how the input "signal" is transmitted down the line, as it travels from one network element to the next. Figure 4.4a suggests that if a "cut" is performed in the ladder immediately following a parallel arm, the input impedance of the remaining properly terminated ladder is equal to Z_0 no matter where the cut is performed. Hence,

$$\frac{v_0}{i_0} = \frac{v_1}{i_1} = \cdots = \frac{v_t}{i_t} = rD_m, \qquad \frac{i_0}{v_1} = \frac{i_1}{v_2} = \cdots = \frac{i_{t-1}}{v_t} = \rho D_m,$$

$$\frac{v_0}{v_1} = \frac{v_1}{v_2} = \cdots = \frac{v_{t-1}}{v_t} = r\rho D_m^2 = \Phi_m^2. \qquad (4.5)$$

From one section to the next, the voltage across a parallel arm is thus divided by Φ_m^2. In a strictly resistive network, such as those investigated so far, Φ_m is a real number > 1, and the network's behavior is that of an *attenuator*.

Examples. 1. It is required to construct an attenuator that divides the input voltage by successive powers of 2. In other words, $v_{i+1}/v_i = 1/\Phi_m^2 = 1/2$. The equation $m = \Phi_m - 1/\Phi_m$ yields $m = 1/\sqrt{2}$, that is, $r\rho = 1/2$. If we arbitrarily pick $r = 1\Omega$, we must have $\rho = 1/2$ mho, the reciprocal of 2Ω. The resulting properly terminated attenuator is shown in figure 4.7a.

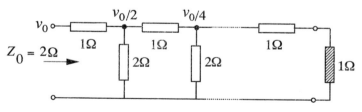

Fig. 4.7a. A powers of 2 attenuator.

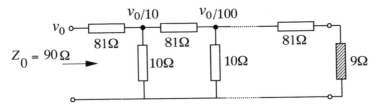

Fig. 4.7b. A powers of 10 attenuator.

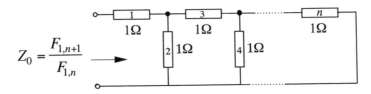

Fig. 4.8. A "Fibonacci ladder."

2. An attenuator that divides the input voltage by successive powers of 10 is such that $m^2 = r\rho = 81/10$. It might have the following characteristics: series arm resistance, $r = 81\Omega$; parallel arm resistance, $1/\rho = 10\Omega$; terminal impedance, 9Ω; input impedance, 90Ω (fig. 4.7b).

3. The network shown in figure 4.8 may be referred to as the *Fibonacci ladder*. All resistances are 1 ohm, and they are numbered 1, 2, 3, ... The input impedance is

$$Z_0 = \frac{F_{1,n+1}}{F_{1,n}} \text{ ohms.}$$

When the number of resistances is sufficiently large, the ladder's input impedance, if measured with sufficient accuracy, is seen to approach 1.618 ohms.

It is often convenient to consider ladder networks made up of a succession of *symmetrical T-sections* as in figure 4.9. It can be easily shown that the ladder's input impedance, when properly terminated, is

$$Z_0 = Z_t = r\left(D_m - \frac{1}{2}\right) = \frac{r}{2}\sqrt{1 + \frac{4}{r\rho}}. \tag{4.6}$$

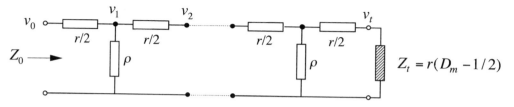

Fig. 4.9. The *T*-section ladder.

Imaginary Components

The ladder networks examined so far were strictly resistive. When we introduced the reader to the notion of continued fractions, we indicated that simple continued fractions (SCFs) could have real as well as complex partial "quotients." Electronic engineers are quite familiar with the three fundamental passive building blocks upon which the edifice of their complex and useful art is predicated. They are the resistance, the inductance, and the capacitance. In a universe in which electrical currents take on various shapes and forms with passing time, the resistor's behavior depends on the instantaneous value of a current, while the inductance's behavior depends on that current's rate of change, and the capacitance's contribution depends on accumulated past events. We owe it to the genius of past mathematicians that such complex behavior, which is normally captured by elaborate equations involving differentiation and integration, was made simple enough for the least mathematically trained engineer to juggle at will, achieved by the bold invention of imaginary quantities, which involve the square root of negative numbers. Without going into excessive detail, suffice it to say that when a *sinusoidal voltage* is applied to a circuit, the remarkably simple Ohm's law can be extended to also encompass the behavior of inductances and capacitances, the so-called *reactive* circuit elements, as follows: If symbol j denotes the square root of -1, and $\omega = 2\pi f$ the *pulsation, or angular velocity*, of a sinusoidal alternating current of frequency f, then the equivalent impedance of inductance L is $j\omega L$, and that of capacitance C is $1/j\omega C$.[1]

---

[1] Extending the notion of impedance to the inductance and capacitance is applicable only to what is referred to as the steady-state solution. Indeed, upon applying a sinusoidal input voltage to a circuit, the onset of current consists of an evanescent transient portion, which dies out exponentially. That stage is followed by a steady-state sinusoidal solution, which can be captured using imaginaries as follows: The fundamental equation relating voltage and current in the case of an inductance L is

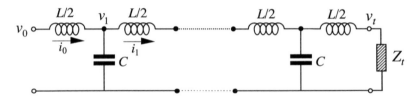

Fig. 4.10a. An inductance-capacitance ladder.

Consider the LC ladder network of figure 4.10a, where each of L and C are considered to be *ideal*, or *lossless* reactances, that is, involving no resistive component capable of internally dissipating electrical energy. The ladder network shown there is properly terminated: both the input and termination impedances are equal to a magnitude referred to as the characteristic impedance, denoted Z_c. Clearly, $m = \sqrt{(jL\omega)(jC\omega)} = j\omega\sqrt{LC}$ is an imaginary number.

Applying equation (4.6), we get

$$Z_c = \frac{j\omega L}{2}\sqrt{1 - \frac{4}{LC\omega^2}} = j\omega L\sqrt{\frac{1}{4} - \frac{1}{LC\omega^2}},$$

and with

$$j\omega L = \sqrt{-L^2\omega^2} = \sqrt{\frac{L}{C}}\sqrt{-LC\omega^2},$$

we get

$$Z_c = \sqrt{\frac{L}{C}}\sqrt{(-LC\omega^2)\left(\frac{1}{4} - \frac{1}{LC\omega^2}\right)},$$

$$Z_c = \sqrt{\frac{L}{C}}\sqrt{1 - \frac{LC\omega^2}{4}}. \tag{4.7}$$

$u = L\,di/dt$. If i is represented by the rotating vector $i = Ie^{j\omega t}$, we get $u = j\omega LIe^{j\omega t} = j\omega Li$, in other words, $u/i = j\omega L$. That is the equivalent of Ohm's law in the case of an inductor. Similarly, the fundamental equation relating voltage and current in the case of a capacitor C is $u = 1/C\int i\,dt$. This yields $u/i = 1/j\omega C$, the equivalent of Ohm's law in the case of a capacitor. Impedances $j\omega L$ and $1/j\omega C$ are referred to as the reactances of inductor L and capacitor C. In chapter VIII, we shall study the transient solution using finite difference methods, which constitute an altogether different approach and a challenging exercise.

We now calculate the value of v_2/v_0, in other words, the voltage following an entire T-section in relation to the input voltage. That ratio is equal to v_{i+2}/v_i for all even values of i if the ladder is properly terminated, that is, $Z_t = Z_c$. In this case the input impedance, as seen from v_0, is also Z_c. Current i_0 is therefore equal to v_0/Z_c, and the voltage drop across the initial inductance is $j\omega L v_0/2Z_c$. Voltage v_1 is equal to v_0 minus the voltage drop:

$$v_1 = v_0 \left(1 - \frac{j\omega L}{2Z_c} \right).$$

The ladder's downstream impedance, as seen from v_2, is again equal to Z_c, from which it follows that the ladder's impedance, as seen from v_1, is $(j\omega L/2) + Z_c$.

Hence,

$$i_1 = \frac{v_1}{\dfrac{j\omega L}{2} + Z_c}$$

and

$$v_2 = i_1 Z_c = \frac{v_1}{\dfrac{j\omega L}{2} + Z_c} Z_c = \frac{v_0 \left(1 - \dfrac{j\omega L}{2Z_c} \right) Z_c}{\dfrac{j\omega L}{2} + Z_c}.$$

This yields

$$\frac{v_2}{v_0} = \frac{2Z_c - j\omega L}{2Z_c + j\omega L}. \tag{4.8a}$$

The Transmission Line

In the simplest case, a transmission line consists of a pair of parallel copper conductors, separated by air. Other elaborate forms may consist of coaxial tubes, filled with an appropriate *dielectric* (nonconducting) material. In a real life lossless transmission line, the capacitive and inductive components are not "lumped" into successive discrete packets

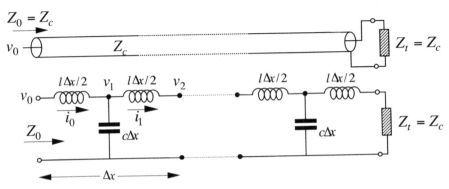

Fig. 4.10b. The properly matched transmission line.

as in figure 4.10a but are evenly distributed along the line as in figure 4.10b. The elementary T-section of length Δx consists of two identical series arms whose inductance is $l\Delta x/2$ and a parallel arm whose capacitance is $c\Delta x$, where l and c respectively denote the line's inductance and capacitance per unit length. Turning to equation (4.7), we may neglect $LC\omega^2 = lc\omega^2(\Delta x)^2$ and write

$$Z_c = \sqrt{\frac{l}{c}}. \tag{4.8b}$$

Z_c is the input impedance of an infinitely long lossless line, or that of an arbitrary finite length of that line, provided it is properly terminated.

The Mismatched Transmission Line

At this point, we may legitimately ask ourselves what happens when the lossless line termination is not the characteristic impedance. In other words, what is the input impedance Z_0 when a line of length x is terminated by an arbitrary impedance Z_t? Neglecting the $j\omega l\Delta x/2$ overhangs at either extremity of the line, we obtain figure 4.10c, which

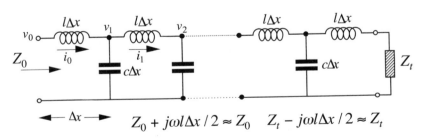

Fig. 4.10c. The transmission line reconfigured.

corresponds exactly to equation (3.39) of chapter III for an improperly terminated simple continued fraction. In that equation, symbol ω is replaced by β to avoid any confusion with the angular velocity, whose universally accepted symbol is ω. (In discussing the dignomonic SPF, we had used symbols α and ω to signify the first and last partial quotients of the corresponding continued fraction):

$$\phi = [\underset{\leftarrow}{\alpha, \alpha', \alpha}, \underset{n}{\ldots}, \underset{\rightarrow}{\beta, \beta'}, \phi_\tau] = \sqrt{\frac{\alpha}{\alpha'}} \left[\underset{\leftarrow}{m, m, m}, \underset{n}{\ldots}, \underset{\rightarrow}{m}, \sqrt{\frac{\beta'}{\beta}} \phi_\tau \right]$$

$$= \sqrt{\frac{\alpha}{\alpha'}} \frac{F_{m,n} + \sqrt{\frac{\beta'}{\beta}} \phi_\tau F_{m,n+1}}{F_{m,n-1} + \sqrt{\frac{\beta'}{\beta}} \phi_\tau F_{m,n}}. \qquad (4.9a)$$

In terms of the transmission line of figure 4.10c, equation (4.9a) corresponds to

$$\alpha = \beta' = j\omega l \Delta x, \qquad \alpha' = \beta = j\omega c \Delta x, \qquad m = j\omega \sqrt{lc} \Delta x,$$
$$Z_0 = \phi, \qquad Z_t = \phi_\tau.$$

In that figure n is an even number, and to every length Δx corresponds a pair of literals m. To integer n thus corresponds a line of length $x = n\Delta x/2$. With m very small,

$$Z_0 = \sqrt{\frac{l}{c}} \frac{F_{m,n \; even} + \sqrt{\frac{c}{l}} Z_t F_{m,n \; odd}}{F_{m,n \; odd} + \sqrt{\frac{c}{l}} Z_t F_{m,n \; even}}. \qquad (4.9b)$$

Applying equations (3.43e) and (3.43f) of chapter III, we obtain

$$Z_0 = \sqrt{\frac{l}{c}} \frac{j \sin(mn\Delta x/2) + \sqrt{\frac{c}{l}} Z_t \cos(mn\Delta x/2)}{\cos(mn\Delta x/2) + \sqrt{\frac{c}{l}} Z_t j \sin(mn\Delta x/2)}$$

$$= Z_c \frac{j\sin(\omega\sqrt{lc}\,x) + \frac{Z_t}{Z_c}\cos(\omega\sqrt{lc}\,x)}{\cos(\omega\sqrt{lc}\,x) + j\frac{Z_t}{Z_c}\sin(\omega\sqrt{lc}\,x)},$$

$$Z_0 = Z_c \frac{jZ_c\sin(\omega\sqrt{lc}\,x) + Z_t\cos(\omega\sqrt{lc}\,x)}{Z_c\cos(\omega\sqrt{lc}\,x) + jZ_t\sin(\omega\sqrt{lc}\,x)}. \tag{4.9c}$$

Obviously, real-life transmission lines are not lossless. The copper conductors have nonzero resistance, and the capacity between conductors is plagued with leakage. It can be shown, in this case, that the purely imaginary constant $j\omega\sqrt{lc}$ has to be replaced by the complex constant $\gamma = \alpha + j\omega\sqrt{lc}$, where the real number α is a function of the resistivity of the conductors and the *leakance* of the dielectric between conductors. It can be also shown that if r is the conductor's resistance per unit length, and g the dielectric's leakance (which has the dimension of admittance) per unit length, then

$$\alpha = \frac{r}{2}\sqrt{\frac{c}{l}} + \frac{g}{2}\sqrt{\frac{l}{c}}.$$

Equation (4.9b) now yields

$$Z_0 = Z_c \frac{Z_c\sinh\gamma x + Z_t\cosh\gamma x}{Z_c\cosh\gamma x + Z_t\sinh\gamma x}. \tag{4.9d}$$

"Telegraphists," as they were referred to a few decades ago, are quite familiar with that equation, which allows them to predict the input impedance of a (short) mismatched transmission line.

Wave Propagation Along a Transmission Line

Returning to the lossless transmission line of figure 4.10b, and using equations (4.8) and (4.9a), we can now calculate the voltage ratio $v_{\Delta x}/v_0$, where Δx is the distance from the origin and consists of an elemental T-section. The method used is rather unorthodox and takes some liberties with strict formalism; nonetheless, it serves our purpose well. Equation (4.8) becomes

$$\frac{v_{\Delta x}}{v_0} = \frac{2\sqrt{\dfrac{l}{c}} - j\omega l\Delta x}{2\sqrt{\dfrac{l}{c}} + j\omega l\Delta x}.$$

Multiplying the numerator and denominator by the latter's conjugate and neglecting second-order infinitesimals, we get

$$\frac{v_{\Delta x}}{v_0} = \frac{4\dfrac{l}{c} - 4j\omega l\sqrt{\dfrac{l}{c}}\,\Delta x}{4\dfrac{l}{c}} = 1 - j(\omega\sqrt{lc})\,\Delta x. \qquad (4.10)$$

If $n\Delta x$ is the distance of abcissa x from the origin, we get

$$v_x = v_{n\Delta x} = v_0(1 - j(\omega\sqrt{lc})\Delta x)^n, \qquad (4.11)$$

and if we put

$$\lim_{\Delta x \to 0}(1 - j(\omega\sqrt{lc})\Delta x) = e^{-j(\omega\sqrt{lc})\Delta x},$$

we get

$$v_x = v_0 e^{-j(\omega\sqrt{lc})n\Delta x} = v_0 e^{-j\omega\sqrt{lc}\,x}. \qquad (4.12)$$

If we now represent the sinusoidal input voltage by the rotating vector[2] $v_0 = Ve^{j\omega t}$, we obtain

$$v_x = Ve^{j\omega(t - \sqrt{lc}\,x)}. \qquad (4.13)$$

That important statement signifies that the vector representing the voltage at abcissa x pulsates at the same frequency f (with the same angular velocity $\omega = 2\pi f$) as the input voltage, but that it lags behind that voltage by an angle θ_x proportional to x. The proportionality coefficient $(\omega\sqrt{lc})$ is known as the transmission line's *propagation constant* at angular velocity ω.

The applied signal thus propagates along the line in the form of a wave that completes a full rotation along the x-axis for every x multiple of λ, where $(\omega\sqrt{lc})\lambda = 2\pi$ that is, $\lambda = 2\pi/\omega\sqrt{lc}$. The quantity λ is the *wavelength*.

[2] Hence the name *angular velocity*.

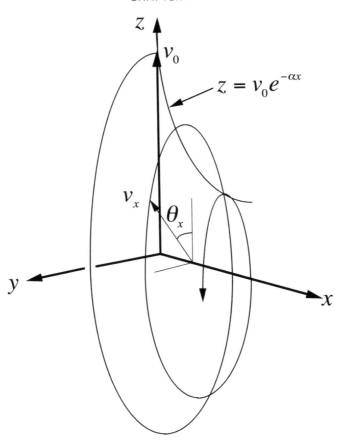

Fig. 4.10d. Wave propagation along the transmission line.

When the line is not lossless, the propagation constant is $\gamma = \alpha + j\omega\sqrt{lc}$, and the wave equation becomes $v_x = Ve^{-\alpha x + j\omega(t - \sqrt{lc}\,x)}$, signifying that the resulting wave progressively dies out exponentially along the x-axis, as shown in figure 4.10d. The figure's projection upon the $y - z$ plane at any given instant is none other than the logarithmic spiral. For every 360° rotation, the spiral's radius decreases by the factor $e^{\alpha\lambda}$.

Seen from any vantage point on the positive x-axis, the spiral rotates counterclockwise at the rate of ω radians per second. That constitutes a perfect metaphor for wave propagation.

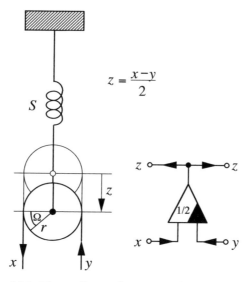

Fig. 4.11. The pulley and its equivalent transducer.

PULLEY LADDER NETWORKS

Consider the pulley in figure 4.11, whose axis is allowed to move longitudinally, as the string wrapped around it is pulled from either side. Spring S is intended to provide only a reaction to the rope's pull and ensure the constuction's stability. In the following discussion, we shall concern ourselves only with longitudinal displacements and rotations, not with forces. If x and y denote longitudinal rope displacements in the respective directions shown, and Ω denotes the rotation of the pulley consecutive to these displacements, let z denote the resulting displacement of the pulley's axis. We can write, with r representing the pulley's radius,

$$z + \Omega r = x \quad \text{and} \quad \Omega r - z = y$$

hence

$$\Omega = \frac{x + y}{2r} \quad \text{and} \quad z = \frac{x - y}{2}$$

The equivalent transducer is represented on the figure's right-hand side.

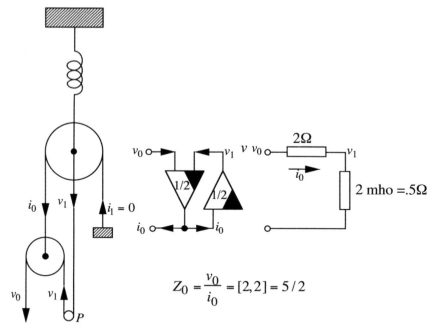

Fig. 4.12. A two-pulley ladder and its equivalent transducer and resistance ladders.

With the help of two such pulleys, we can construct the machine shown in figure 4.12, in which an auxiliary fixed axis pulley P plays no part in the network other than reversing the direction of displacement v_1. The equivalent transducer and electrical networks are shown as well. Note that the statement $i_1 = 0$ corresponds in the pulley network to immobilizing the corresponding string's extremity, and in the electrical circuit to leaving the network open, where the subsequent series arm should have been attached. To these constructions corresponds equation

$$\frac{v_0}{i_0} = [2, 2] = \frac{5}{2}.$$

Arrows indicate displacements in the mechanical model, "signals" in the transducer model, and currents in the electrical model. The next construction, shown in figure 4.13, involves three pulleys. The process can be continued, involving more and more pulleys. The input impedance of an n-pulley construction is $v_0/i_0 = F_{2,n+1}/F_{2,n}$.

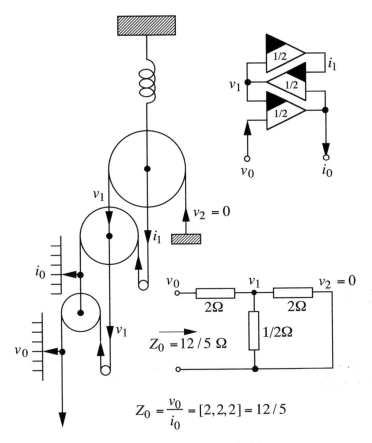

$$Z_0 = \frac{v_0}{i_0} = [2,2,2] = 12/5$$

Fig. 4.13. A three-pulley ladder.

None of the networks constructed so far was properly terminated. The "characteristic impedance" of these networks is $Z_c = 1 + \sqrt{2}$. That would also be the value of the input impedance, should the ladders be properly terminated. Figure 4.14 shows a properly terminated network, corresponding to the continued fraction $Z_0 = Z_c = [2, 2, 2, 2, Z_c] = 1 + \sqrt{2}$. Its termination consists of two integrally joined pulleys of radii 1 and $1 + \sqrt{2}$, whose common axis is fixed, and around which the string's overhanging extremities are wrapped. In this construction, $v_0/i_0 = v_2/i_2 = 1 + \sqrt{2}$.

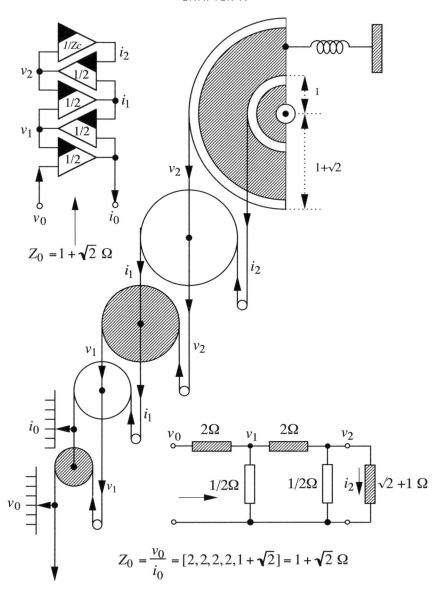

$$Z_0 = 1 + \sqrt{2}\ \Omega$$

$$Z_0 = \frac{v_0}{i_0} = [2, 2, 2, 2, 1 + \sqrt{2}] = 1 + \sqrt{2}\ \Omega$$

Fig. 4.14. A properly terminated pulley ladder and its equivalent transducer and resistance ladders.

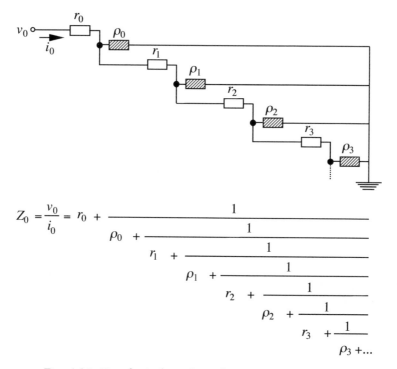

$$Z_0 = \frac{v_0}{i_0} = r_0 + \cfrac{1}{\rho_0 + \cfrac{1}{r_1 + \cfrac{1}{\rho_1 + \cfrac{1}{r_2 + \cfrac{1}{\rho_2 + \cfrac{1}{r_3 + \cfrac{1}{\rho_3 + \dots}}}}}}}$$

Fig. 4.15. Topological similarity between the ladder and its continued fraction.

MARGINALIA

A Topological Similarity

Figure 4.15 highlights a striking *topological* similarity between the electrical ladder network and the continued fraction that mathematically describes its behavior. A detailed examination of the analogy provides useful insight into the physical process involved, as well as the adequacy of its mathematical representation. Starting at the bottom end of the continued fraction, the nature of the successive retrovergents, or more precisely their dimension, alternates between that of admittance (hatched in the upper diagram) and that of impedance, whereas the successive convergents all have the dimension of impedance. That metaphor is perhaps still more striking in figure 4.14.

Whorled Figures

In this chapter, we shall study a geometric metaphor of periodic continued fractions, which leads to the equiangular as well as other spirals described in subsequent chapters. The study reveals interesting gnomonic figures such as the Fibonacci whorl, the homognomonic rectangle and triangle, and others.

WHORLED RECTANGLES

Given positive numbers $a_0 > a_1$, the procedure known as Euclid's algorithm, which is aimed at obtaining their greatest common denominator, was described in chapter 2. It is restated here:

$$a_0 = a_1 q_0 + a_2, \quad a_1 > a_2,$$
$$a_1 = a_2 q_1 + a_3, \quad a_2 > a_3,$$
$$a_2 = a_3 q_2 + a_4, \quad a_3 > a_4,$$
$$a_3 = a_4 a_3 + a_5, \quad a_4 > a_5,$$
$$\cdot \ \cdot \ \cdot \ \cdot \ \cdot$$
$$a_{i-1} = a_i q_{i-1} + a_{i+1}, \quad a_i > a_{i+1},$$
$$\cdot \ \cdot \ \cdot \ \cdot \ \cdot$$
$$a_{n-1} = a_n q_{n-1} + 0,$$

where $q_0, q_1, q_2, \ldots$ are integers.

Euclid's Algorithm

Let us now turn to figure 5.1 and consider rectangle $ABCD$ whose sides are a_0, a_1. Rectangle $CBEF$ of sides a_0, $a_1 q_0$ is inscribed inside $ABCD$ and shares side BC with it. Side AE of the remaining rectangle $AEFD$ is equal to a_2. Rectangle $AEHG$ is now inscribed inside $AEFD$, with its vertical side AG equal to $a_2 q_1$. The remainder GD is equal to a_3. That process, which may be extrapolated without difficulty, constitutes a geometric metaphor for the above equations.

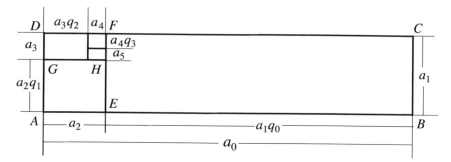

Fig. 5.1. A geometric metaphor for the equations of Euclid's algorithm.

In figure 5.2, rectangle $ABCD$ has sides 43 and 9. It is said to be of *horizontal proportion* or *h*-proportion 43/9. Beginning at its far right, as many adjacent squares are inscribed as possible. They turn out to be four in number, covering area $EBCF$. That leaves residual rectangle $AEFD$ of sides 9 and 7, which is said to be of *vertical proportion*, or *v*-proportion 9/7. So far, the construction is the geometric equivalent of $43 = 4 \times 9 + 7$, or

$$\frac{43}{9} = 4 + \frac{1}{9/7},$$

that is,

$$p_3 = \frac{43}{9} = [4, 9/7] = [4, p_2].$$

The next step consists in inscribing as many adjacent squares as possible

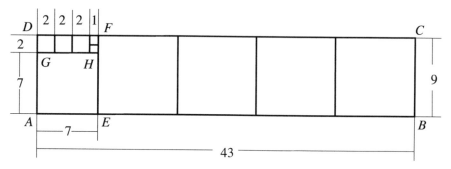

Fig. 5.2. The whorled rectangle of ratio 43/9.

in the residual rectangle, starting from the bottom. Only one such square, namely *AEHG,* can be drawn, leaving residual rectangle *GHFD* of sides 7 and 2 (of *h*-proportion 7/2) at the top. This is equivalent to

$$\rho_2 = \frac{9}{7} = 1 + \frac{7}{2}$$

that is,

$$\frac{43}{9} = [4, 1, 7/2] = [4, 1, \rho_1].$$

The third step is equivalent to

$$\rho_1 = \frac{7}{2} = 3 + \frac{1}{2}$$

and leaves a rectangle of *v*-proportion 2/1. The process halts at that point, as 2 is an integer, and the following continued fraction can be constructed:

$$\frac{43}{9} = [4, 1, 3, \rho_0] = [4, 1, 3, 2].$$

Note that quotient 2 may in turn be expressed as $2 = 1 + (1/1)$, yielding

$$\frac{43}{9} = [4, 1, 3, 1, 1].$$

Observe the change in direction from right to left and from bottom to top, following a clockwise *inward whorl*. The rectangles of figures 5.1 to 5.3 will be referred to as *whorled rectangles*. They are the precursors of spirals, which will be discussed in chapter VIII. The same figure could have been obtained, starting with an inner square seed, by adding to it a sequence of rectangles of proportions 1, 3, 1, 4, following an *outward* whorl. We now examine what happens when the continued fraction is periodic.

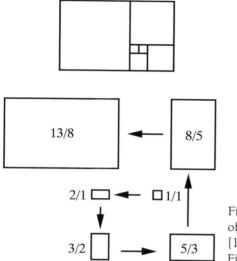

13/8 ⟵ 8/5

2/1 ⟵ 1/1

3/2 ⟶ 5/3

Fig. 5.3. Successive generations of rectangles corresponding to [1, 1, 1, 1, 1, 1] or the order 1 Fibonacci whorl.

Monognomonic Whorled Rectangles

The outward whorl of figure 5.3 corresponds to the continued fraction [1, 1, 1, 1, 1, 1]. Starting with an inner seed of proportion $\rho_0 = 1$ and adding a square to its left results in a rectangle of h-proportion $\rho_1 = 2/1$. As the outward whorl proceeds, the successive proportions are 1/1, 2/1, 3/2, 5/3, 8/5, 13/8, corresponding to the continued fraction's retrovergents. The sequence 1, 1, 2, 3, 5, 8, 13 appearing in both numerators and denominators is none other than the Fibonacci sequence of order 1.

Figure 5.4 corresponds to the continued fraction [2, 2, 2, 2], and its successive retrovergents are 2/1, 5/2, 12/5, 29/12. The sequence 1, 2, 5, 12, 29 is the Fibonacci sequence of order 2. The previous two

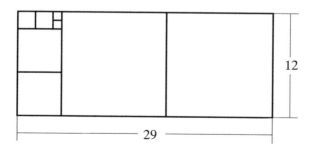

12

29

Fig. 5.4. The whorled rectangle [2, 2, 2, 2] or the order 2 Fibonacci whorl.

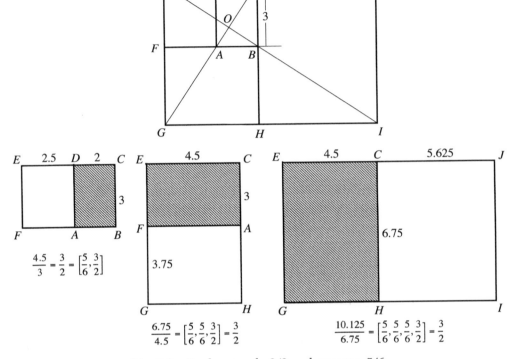

Fig. 5.5a. Seed rectangle 3/2 and gnomon 5/6.

examples correspond to *improperly terminated* (seeded) SPFs. Their retrovergents differ from one to the next. We now turn our attention to *properly terminated* (seeded) SPFs. We shall first address, as an example, the properly terminated (seeded) SPF whose partial quotient is 5/6 and seed 3/2 (fig. 5.5a):

$$\Phi_{5/6} = \frac{3}{2} = \left[\frac{5}{6}, \frac{5}{6}, \frac{5}{6}, \dots, \frac{3}{2} \right].$$

Starting with an inner seed *ABCD* of v-proportion 3/2, we add to its left the rectangle *ADEF* of h-proportion 5/6. Its height is 2.5 and its base 3. The resulting rectangle *BCEF* is of h-proportion 3/2, similar to the seed. As we continue the outward whorl, every addition of a new

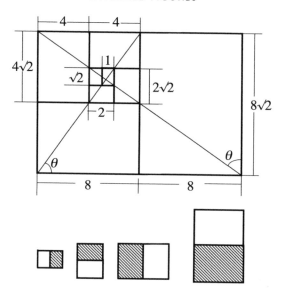

Seed rectangle √2 and gnomon 1/√2

Fig. 5.5b. The homognomonic rectangle.

gnomon of proportion 5/6 results in a rectangle similar to the seed, albeit rotated by 90°. We next address another properly seeded simple periodic continued fraction (SPF) whose partial quotient is $1/\sqrt{2}$, and seed $\sqrt{2}$ (fig. 5.5b):

$$\Phi_{1/\sqrt{2}} = \sqrt{2} = \left[\frac{1}{\sqrt{2}}, \frac{1}{\sqrt{2}}, \ldots, \frac{1}{\sqrt{2}}, \sqrt{2} \right].$$

Starting with an inner seed of v-proportion $\sqrt{2}$ (fig. 5.5b), we add an identical rectangle to its left. As we continue the outward whorl, each rectangle is doubled, and its proportion (vertical or horizontal) is $\sqrt{2}$. The rectangle of proportion $\sqrt{2}$ is therefore its own gnomon. The figure is said to be *homognomonic*. Irrational number $\sqrt{2}$ is approximately the proportion of a standard sheet of stationary in France, whose official format is 29.7 × 21 centimeters. The very remarkable advantage of that choice is that, no matter how many times you halve the sheet of paper, every new half is exactly in the same proportion, namely, $\sqrt{2}$. Each new sheet, folio, quarto, or octavo, is in the same proportion as the original, resulting in pleasing harmony, and perhaps economic gain.

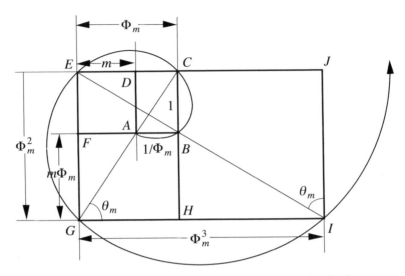

Fig. 5.6. Rectangular monognomonic outward whorl.

The German photographic DIN format is predicated upon that property of the homognomonic rectangle.

Figure 5.6 shows the general appearance of an outward whorled rectangle of seed Φ_m and gnomon m. The diagonals intersect at right angles, and we have $\tan \theta_m = \Phi_m$. Observe that no such diagonals can be drawn for improperly terminated SPFs, such as those of figure 5.3. That property of monognomonic whorled rectangles can be easily deduced using elementary geometry. *Clearly, a whorled figure may be referred to as monognomonic only inasmuch as it is properly seeded (or infinite).* The properties of its spiral envelope will be studied in a subsequent chapter.

Dignomonic Whorled Rectangles

In figures 5.7a and 5.7b, *ABCD* is inscribed in *BCEF* and shares side *BC*. Four diagonals can be drawn and grouped in pairs, one diagonal from each primary rectangle. There are four such pairs. Of these, two are such that the diagonals intersect at a rectangle's vertex. The remaining two pairs are such that the diagonals intersect somewhere inside the inscribed rectangle. We shall only consider the latter two pairs, namely *AC*, *BE* of figure 5.7a, and *BD*, *CF* of figure 5.7b. Intersec-

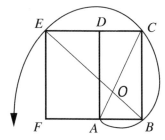

Fig. 5.7a. Left-handed whorl.

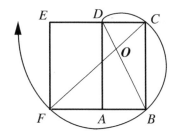

Fig. 5.7b. Right-handed whorl.

tion O is called the origin or pole. Of the six vertices in the diagram, only four fall on the selected diagonals. They are called diagonal vertices, and the straight line segment joining pole O to a diagonal vertex is called a radius.

Starting from the diagonal vertex closest to origin O, namely, A in figure 5.7a, and visiting the other three diagonal vertices in order of increasing radius length, we describe an expanding spiral $ABCE$, and our excursion takes place counterclockwise. A spiral which expands anticlockwise is said to be left-handed. Had we started at D in figure 5.7b, our excursion along the expanding spiral would have taken place clockwise. It is a mirror image of the other spiral, albeit rotated 180° and is said to be right-handed.

Turning to figure 5.7c, any line segment joining two successive diagonal vertices will be referred to as a *chord*. Proceeding counterclockwise from A, we measure successive chords AB, BC, CE, and put $BC/AB = \phi_s$ and $CE/BC = \phi_r$. Here ϕ_s is the v-proportion of $ABCD$, and ϕ_r is the h-proportion of $BCEF$. As the spiral expands, more and more rectangles make their appearance. It is therefore necessary to agree upon a rule defining a rectangle's proportion. Any rectangle, two of whose adjacent sides are successive chords of the spiral, will be called *primary*. Its leading chord, as the spiral expands, is called the *height*, and its trailing chord the *base*. The rectangle's proportion may then be defined as the ratio height/base. That proportion is horizontal when the rectangle's height is horizontal, and vertical otherwise. We can now equate the length of common side BC to 1, making $AB = 1/\phi_s$ and $CE = \phi_r$. The numbers r and s are defined as

$$r = \phi_r - \frac{1}{\phi_s}, \qquad s = \phi_s - \frac{1}{\phi_r}. \qquad (5.1)$$

Whereas the number r is seen to correspond to the length of segment

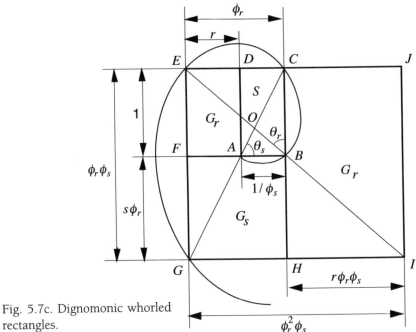

Fig. 5.7c. Dignomonic whorled
rectangles.

DE (actually to the ratio DE/BC), the number s is somewhat less geometrically obvious at this stage. As we shall be dealing only with real
positive numbers r, ϕ_r, ϕ_s, it follows that

$$\phi_r \phi_s > 1,$$ (5.2)

$$s\phi_r = r\phi_s = \phi_r \phi_s - 1,$$ (5.3)

and s is thus also a positive number. Imagine that the plane of figure
5.8a, which represents a monognomonic whorled rectangle of seed

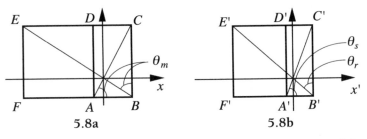

Fig. 5.8. The dignomonic whorled rectangle as a projection of the
monognomonic rectangle.

Φ_m, is tilted around axis Oy by some angle a with respect to the horizontal plane, and that the spiral is projected perpendicularly upon that plane, as in figure 5.8b. We obtain the following relationships:

$$A'B' = AB \cos a, \qquad C'E' = CE \cos a,$$

$$\phi_r = \frac{C'E'}{A'B'} = \frac{CE \cos a}{1} = \Phi_m \cos a,$$

$$\phi_s = \frac{C'B'}{A'B'} = \frac{1}{AB \cos a} = \frac{\Phi_m}{\cos a},$$

$$m = DE, \qquad r = DE \cos a, \qquad s = \frac{DE}{\cos a},$$

$$\varphi_r \phi_s = \Phi_m^2, \qquad \frac{\phi_r}{\phi_s} = \frac{r}{s} = \cos^2 a.$$

The dignomonic spiral results from a horizontal contraction of the monognomonic spiral and angle $AOB \leq \pi/2$. A rotation around the x-axis would result in a spiral that is contacted in the vertical direction.

Returning to figure 5.7c, we now extend side EF downwards until it intersects diagonal CA at G and complete rectangle $CEGH$. We have

$$\frac{EG}{DA} = \frac{CE}{CD} = \phi_r \phi_s,$$

hence

$$FG = \phi_r \phi_s - 1 = s\phi_r = s \times CE$$

and

$$s = \frac{FG}{CE}.$$

We observe the following, starting with seed S:

Rectangle $R = (S + G_r)$ of h-proportion ϕ_r = (seed rectangle S of v-proportion ϕ_s) + (rectangle G_r of h-proportion ϕ_r)

Rectangle $(S + G_r + G_s)$ of v-proportion ϕ_s = (rectangle $S + G_r$ of h-proportion ϕ_r) + (rectangle G_s of v-proportion ϕ_s)

Rectangle $(S + G_r + G_s + G_{r'})$ of h-proportion ϕ_r = (rectangle $S + G_r + G_s$ of v-proportion ϕ_s) + (rectangle $G_{r'}$ of h-proportion ϕ_r).

The above process can be continued indefinitely, generating rectangles of alternating proportions ϕ_r and ϕ_s, thanks to the addition of G_r, G_s, $G_{r'}$ .... The resulting spiral is wrapped around the successive primary rectangles, connecting their successive diagonal vertices. Returning to equations (5.1), we can write

$$\phi_r = r + \frac{1}{\phi_s}, \qquad \phi_s = s + \frac{1}{\phi_r}. \qquad (5.4)$$

This generates the periodic continued fraction

$$\phi_r = [r, s, r, s, \ldots, r, \phi_s], \qquad (5.5)$$

which corresponds to a seed rectangle of v-proportion ϕ_s, or

$$\phi_r = [r, s, r, s, \ldots, s, \phi_r], \qquad (5.6)$$

corresponding to a seed rectangle of h-proportion ϕ_r. Similarly, we can write

$$\phi_s = [s, r, s, r, \ldots, r, \phi_s], \qquad (5.7)$$

$$\phi_s = [s, r, s, r, \ldots, s, \phi_r]. \qquad (5.8)$$

The reader will have recognized the statements of chapter III concerning dignomonic continued fractions, where $\alpha = r$, $\alpha' = s$, $\phi_{\alpha,\alpha'} = \phi_r$, $\phi_{\alpha',\alpha} = \phi_s$.

Statement $\phi_s = [s, \phi_r]$ can be geometrically interpreted as follows: Vertically adding to a primary rectangle of h-proportion ϕ_r a rectangle of v-proportion s results in a new primary rectangle of v-proportion ϕ_s. The same statement can be repeated, swapping the adjectives vertical and horizontal, along with literals r and s. Generating the next rectangle, of h-proportion ϕ_r corresponds to $\phi_r = [r, s, \phi_r] = [r, \phi_s]$. The L-shaped form Γ_s of figure 5.9a, consisting of rectangles G_r and G_s taken

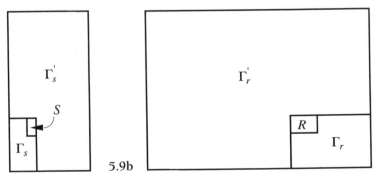

5.9a   5.9b

Fig. 5.9. Gnomons and semignomons.

together, is the *gnomon* of S, as its addition to S generates a geometrically similar rectangle. We can thus refer to rectangles G_s and G_r of figure 5.7c as *semignomons*. Gnomon Γ'_s is equal to gnomon Γ_s enlarged by a factor $\phi_r\phi_s$. Similarly, the L-shaped form Γ_r of figure 5.9b consisting of rectangles G_s and G'_r taken together, is the gnomon of R, where $R = S + G_r$. Gnomon Γ'_r is equal to gnomon Γ_r enlarged by a factor $\phi_r\phi_s$.

Figure 5.10a represents a dignomonic whorled rectangle drawn for $r = 1, s = 1/2$. In figure 5.10b, the seed is the inner square of size 1, whereas in figure 5.10c, it is the inner rectangle of size 1×2.

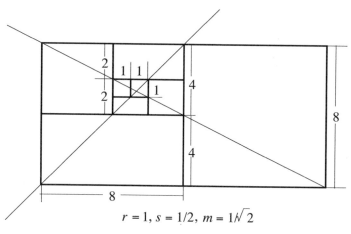

$$r = 1, s = 1/2, m = 1\sqrt{2}$$

Fig. 5.10a. Dignomonic whorled rectangle with $r = 1, s = 1/2$.

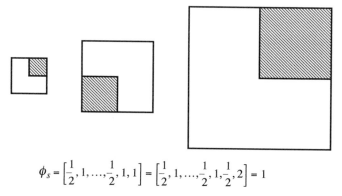

$$\phi_s = \left[\frac{1}{2}, 1, ..., \frac{1}{2}, 1, 1\right] = \left[\frac{1}{2}, 1, ..., \frac{1}{2}, 1, \frac{1}{2}, 2\right] = 1$$

Fig. 5.10b. The construction of fig. 5.10a with the inner square as seed.

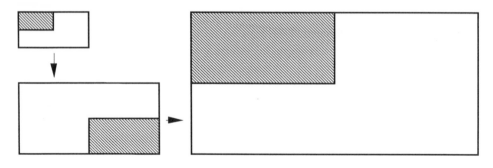

$$\phi_r = \left[1, \frac{1}{2}, \ldots, 1, \frac{1}{2}, 2\right] = \left[1, \frac{1}{2}, \ldots, 1, \frac{1}{2}, 1, 1\right] = 2$$

Fig. 5.10c. The construction of fig. 5.10a with inner rectangle as seed.

Whereas a rectangular shape of proportion $\Phi_m > 1$ possesses a nonzero gnomon $m = \Phi_m - 1/\Phi_m$, the square has no single *monognomon* but can receive an infinity of *dignomons*, all of which satisfy $r - s - rs = 0$.

SELF-SIMILARITY

Consider the properly terminated dignomonic SPF

$$\phi_{\alpha,\alpha'} = [\alpha, \alpha', \ldots, \omega, \omega', \phi_{\omega,\omega'}]. \tag{5.9a}$$

It can be written

$$\phi_{\alpha,\alpha'} = [\overset{\longleftarrow \ p \ \longrightarrow}{\alpha, \alpha', \alpha, \ldots, \mu, \mu'}, [\overset{\longleftarrow \ q \ \longrightarrow}{\mu, \mu', \ldots, \omega, \omega', \phi_{\omega,\omega'}}]] \tag{5.9b}$$

$$= [\alpha, \alpha', \ldots, \mu, \mu', \phi_\mu], \tag{5.9c}$$

where each of generic symbols μ, μ' represent either α or α'. The geometric interpretation of these statements is the following.

Starting with seed $\phi_{\omega,\omega'}$, and following $p + q$ iterations, a rectangle of proportion $\phi_{\alpha,\alpha'}$ is reached. Starting with the same seed, and following q iterations, a rectangle of proportion $\phi_{\mu,\mu'}$ is reached, which becomes the seed for an additional p iterations, also leading to $\phi_{\alpha,\alpha'}$. Expressions (5.9) are properly terminated, meaning that any primary rectangle encountered in the course of a construction can be regarded as a new seed, and the construction can proceed in either direction from that arbitrarily chosen seed, generating an expanding or contracting spiral. That process is infinite in both directions, as $\phi_{\alpha,\alpha'}$ can itself be looked upon as a seed. Imagine that you already proceeded with a sufficient number of iterations in the contracting or in the expanding direction. Imagine also that you possess a photocopying machine that can enlarge or reduce by a ratio exactly equal to $\phi_{\alpha,\alpha'}\,\phi_{\alpha',\alpha} = \Phi_m^2$. If you enlarge or reduce your figure any number of times by that ratio, and print it on transparent paper, then apply it above the original figure, you will observe that the two figures exactly coincide, albeit with overhangs on either side, since your original drawing was by necessity finite. Had your drawing been infinite in both directions, the coincidence would have been total, and the figures indistinguishable from one another.

In the case of a monognomonic whorled rectangle, enlarging or contracting the figure by a factor of Φ_m causes the coincidence of the figures with every 90° rotation in either direction. That character of self-similarity is the very essence of spirals and has fascinated generations of mathematicians, including Bernoulli, who deemed the curve miraculous.

IMPROPERLY SEEDED WHORLED RECTANGLES

Despite the semantic ambiguities that our terminology might entail, the *seed* rectangle's proportion is actually the continued fraction's *termination*. The notion of termination is inspired by the physical impedance that is placed at the far end of an electrical network or a transmission line, whereas the notion of seed is inspired by plant and animal growth. We have seen that for large values of n, regardless of termination ϕ_τ,

$$\phi = [\alpha, \alpha', \alpha, \ldots, \omega, \omega', \phi_\tau] \xrightarrow[n \text{ large}]{} \sqrt{\frac{\alpha}{\alpha'}}\,\Phi_m = \phi_{\alpha,\alpha'}\,.$$

This suggests that if we start a construction with an arbitrary seed of proportion ϕ_τ, every iteration brings the rectangle's proportion closer and closer to $\phi_{\alpha,\alpha'}$ or $\phi_{\alpha',\alpha}$ depending on whether the number of iterations is even or odd. Figure 5.11a shows a dignomonic outward whorl, with $\alpha = 0.07$, $\alpha' = 0.035$ and a seed rectangle of proportion $\phi_\tau = 10$. To these values corresponds $D \approx 20.709237$, hence $\phi_{\alpha,\alpha'} \approx 1.4496466$ and $\phi_{\alpha',\alpha} \approx 0.7248233$. These are the proportions of

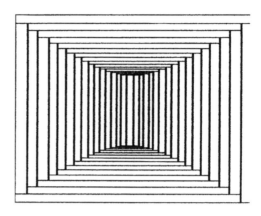

Fig. 5.11a. Improperly seeded dignomonic whorled rectangles.

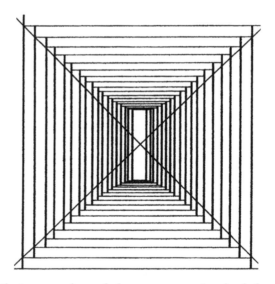

Fig. 5.11b. Improperly seeded monognomonic whorled rectangles.

the two outermost rectangles, to which the successive primary rectangles gracefully tend. Figure 5.11b was obtained for a monognomonic outward whorl, with $\phi_\tau = 5$, $m = 0.05$, corresponding to $\Phi_m \approx$ 1.02531245. If the form is iterated several times, both outermost primary rectangles' proportions approach that value. The reader may reflect upon the orthogonality of the asymptotes in the latter case.

Two Whorled Triangles

Obviously, rectangles are not the only geometric figures that can be whorled. Figure 5.12a shows the whorled equilateral triangle, a monognomonic figure. The chord expands by a factor 2 for every 60° rotation, whereas the radius extending from pole O to vertices A, B, C, D, . . . expands by the same factor for every 120° rotation. The gnomon

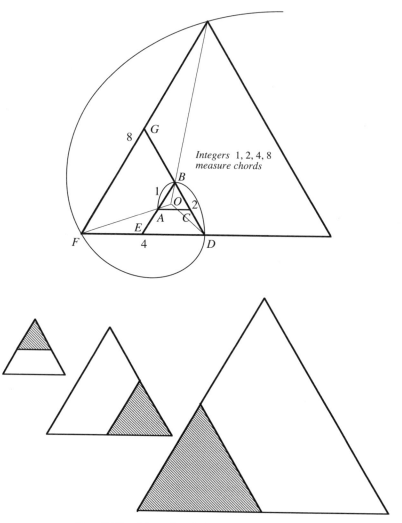

Fig. 5.12a. The whorled equilateral triangle.

of equilateral triangle *ABC* is the trapezoid *ACDE'*. Figure 5.12b shows the whorled isoceles right triangle. The triangle is identical to its gnomon and can thus be referred to as homognomonic. Two distinct configurations can be identified, namely *ABCDE* and *A'B'C'D'E'*. In both instances, the chord doubles for every 90° rotation. After studying the gold and silver numbers, we will examine other whorled polygons.

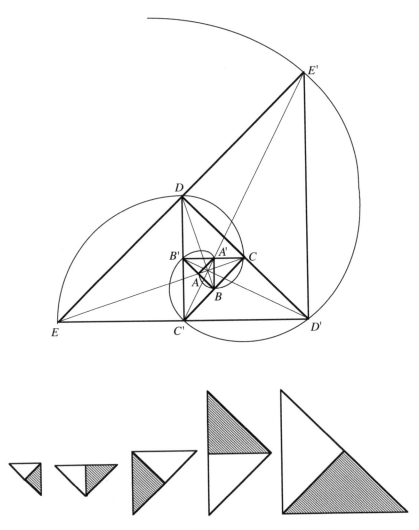

Fig. 5.12b. The whorled homognomonic triangle.

MARGINALIA

Transmission Lines Revisited

The whorled figures of figure 5.11b are reminiscent of the electrical transmission lines described in chapter IV. If a transmission line is terminated by a load exactly equal to its *characteristic impedance* Z_c, the line's impedance viewed from the input end is also equal to that impedance, and the line is said to be *matched* or *properly terminated* (fig. 5.13a). The statement is akin to $\Phi_m = [\underset{\xleftarrow{\hspace{1em}} n \xrightarrow{\hspace{1em}}}{m, m, \ldots, m, \Phi_m}]$. If the load impedance Z_τ is different from the characteristic impedance, its input impedance Z_0 is different from both impedances, and the line is said to be *mismatched* or *improperly terminated*. Regardless of the load impedance, the input impedance gradually approaches the characteristic impedance, as the line increases in length (fig. 5.13b). That, in turn, is akin to $[\underset{\xleftarrow{\hspace{1em}} n \xrightarrow{\hspace{1em}}}{m, m, m, \ldots, m}] \xrightarrow[n \to \infty]{} \Phi_m$, and the input impedance of an infinitely long transmission line is equal to its characteristic impedance (fig. 5.13c).

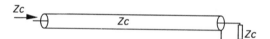

Fig. 5.13a. A properly terminated transmission line.

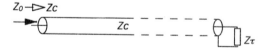

Fig. 5.13b. An improperly terminated transmission line.

Fig. 5.13c. An infinite transmission line.

The Golden Number

Geometry conceals two great treasures: One is the theorem
of Pythagoras, the other the division of a line into middle
and extreme ratio. The first is comparable to a measure
of gold, and the second to a precious jewel.
(*Johannes Kepler*)

A number that has fascinated mathematicians, architects, schol-
ars, painters, musicians, and naturalists for centuries is the positive
root of the equation

$$\Phi_1^2 - \Phi_1 - 1 = 0, \tag{6.1a}$$

namely,

$$\Phi_1 = \frac{1 + \sqrt{5}}{2} \approx 1.618033989. \tag{6.1b}$$

It is known as the *golden ratio,* or *divine proportion,* or *golden section.*
In the following, we shall use the symbol ϕ instead of Φ_1 to denote
the gold number, consistent with mathematical literature. The letter ϕ
is the initial of Phidias, who adorned the Parthenon with his spendid
sculptures. Legend has it, ironically, that Phidias was driven by his
excessive love of gold to steal large amounts of the precious metal,
which caused him to die miserably in prison.
From equation (6.1a) it follows that

$$\phi = \sqrt{1 + \phi} = \sqrt{1 + \sqrt{1 + \phi}}$$
$$= \sqrt{1 + \sqrt{1 + \sqrt{1 + \phi}}} = \dots, \tag{6.2a}$$

and it can be shown that expression (6.2b) converges to ϕ:

$$\sqrt{1 + \sqrt{1 + \sqrt{1 + \sqrt{1 + \ldots}}}} \to \phi. \qquad (6.2b)$$

Similarly,

$$\sqrt[3]{1 + 2\sqrt[3]{1 + 2\sqrt[3]{1 + 2\sqrt[3]{\ldots}}}} \to \phi. \qquad (6.2c)$$

Continued fractions (6.3a) and (6.3b) exhibit similar convergence properties:

$$\phi = 1 + \frac{1}{\phi} = 1 + \cfrac{1}{1 + \cfrac{1}{\phi}} = 1 + \cfrac{1}{1 + \cfrac{1}{1 + \cfrac{1}{\phi}}} = \ldots \qquad (6.3a)$$

$$1 + \cfrac{1}{1 + \cfrac{1}{1 + \cfrac{1}{1 + \cfrac{1}{\ldots}}}} \to \phi. \qquad (6.3b)$$

Consider the following sequence, which stretches to infinity in both directions:

$$\ldots, \frac{1}{\phi^3}, \frac{1}{\phi^2}, \frac{1}{\phi}, 1, \phi, \phi^2, \phi,^3 \ldots \qquad (6.4a)$$

Substituting $\phi^2 = 1 + \phi$, the sequence can be rewritten

$$\ldots, (2\phi - 3), (2 - \phi), (\phi - 1), 1, \phi, (1 + \phi),$$
$$(1 + 2\phi), (2 + 3\phi), \ldots \qquad (6.4b)$$

The integral coefficients in the above sequence are none other than the terms F_i of the Fibonacci sequence, where $F_0 = 0$, $F_1 = 1$, and

TABLE 6.1

Values of $F_{1,i} = F_i$ for $i = -3$ to 11

i	−3	−2	−1	0	1	2	3	4	5	6	7	8	9	10	11
F_i	2	−1	1	0	1	1	2	3	5	8	13	21	34	55	89

$F_{i+1} = F_{i-1} + F_i$, as in table 6.1. The general term of sequence (6.4b) is easily found, by inspection, to be

$$\phi^i = \phi F_i + F_{i-1}, \qquad (6.5a)$$

a result that may be proven by induction over i, from which it follows that ϕ is the real positive solution of equation

$$x^i - F_i x - F_{i-1} = 0, \qquad (6.5b)$$

for all values of i. For example,

$$x^3 - 2x - 1 = 0, \qquad x^5 - 5x - 3 = 0.$$

Each term of the sequences (6.4a) and (6.4b) is the sum of the two preceding terms, and the ratio of any two consecutive terms is constant. That unique sequence is usually referred to as the *golden sequence*. We can write

$$\phi = \frac{1+\phi}{\phi} = \frac{1+2\phi}{1+\phi} = \frac{2+3\phi}{1+2\phi} = \frac{3+5\phi}{2+3\phi} = \frac{5+8\phi}{3+5\phi} = \cdots \qquad (6.5c)$$

Whereas equation (6.5a) expresses ϕ^i in terms of Fibonacci numbers, equation (6.6a), which was initially derived upon studying Fibonacci sequences, achieves exactly the converse:

$$F_i = \frac{\varphi^i - \left(-\dfrac{1}{\varphi}\right)^i}{\phi + \dfrac{1}{\phi}} = \frac{\varphi^i - (1-\varphi)^i}{2\phi - 1}, \qquad (6.6a)$$

from which the famous expression discovered by de Moivre in 1718, can be derived

$$F_i = \frac{1}{\sqrt{5}} \left(\left(\frac{1 + \sqrt{5}}{2} \right)^i - \left(\frac{1 - \sqrt{5}}{2} \right)^i \right). \tag{6.6b}$$

For large values of i, equation (6.6a) yields

$$F_i \approx \frac{\phi^i}{2\phi - 1}, \tag{6.6c}$$

and

$$\phi \approx \frac{F_{i+1}}{F_i}, \tag{6.6d}$$

as well as

$$\phi^i \approx 2F_{i+1} - F_i. \tag{6.6e}$$

For example,

$(6.6c) \rightarrow$
$$\frac{\phi^{10}}{2\phi - 1} \approx \frac{122.9918693}{2.236067977} \approx 55.00363368, \quad \text{whereas } F_{10} = 55,$$

$(6.6d) \rightarrow$
$$\frac{F_{11}}{F_{10}} = \frac{89}{55} = 1.6181818\ldots \quad \text{and} \quad \frac{F_{25}}{F_{24}} = \frac{75,025}{46,368} \approx 1.618033989,$$

$(6.6e) \rightarrow$
$$2F_{11} - F_{10} = 2 \times 89 - 55 = 123, \quad \text{whereas } \phi^{10} \approx 122.9918693.$$

FROM NUMBER TO GEOMETRY

Figure 6.1 shows one of several methods that can be used for drawing a rectangle whose proportion is the golden ratio, using ruler and compass. Starting with straight segments OA, OF of unit length, right triangle OAD is drawn, with side AD = 2. Its hypotenuse OD

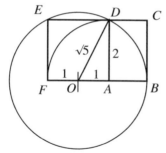

Fig. 6.1. Constructing a golden rectangle.

measures $\sqrt{5}$ units. Drawing a circle of center O and radius OD allows the construction of rectangles $ABCD$ and $BCEF$, both of which are of aspect ratio (ratio of long to small side) equal to the golden ratio.

If a square gnomon is erected upon the long side of a divine rectangle, the resulting rectangle is itself divine. In other words,

$$\frac{length}{width} = \frac{length + width}{length}.$$

If a line segment BF (such as that of fig. 6.1), is divided into segments AB and AF, in such a manner that the *"middle section"* AF/AB is equal to ϕ, it follows that the *"extreme section"* BF/AF is also equal to ϕ, hence the name golden *section*. That is equivalent to the statement that the golden number's mantissa $\phi' = 0.6180339 \ldots$ is such that $1 + \phi' = 1/\phi'$

THE WHORLED GOLDEN RECTANGLE

Figure 6.2a corresponds to the left-handed[1] whorl starting with an inner *seed* rectangle $ABCD$ of vertical proportion ϕ. Only the initial steps of an infinite process are shown, which it would obviously be in vain to attempt to represent on paper. Equation (6.5c) corresponds to

$$\frac{BC}{AB} = \frac{CE}{BC} = \frac{EG}{CE} = \frac{GI}{EG} = \phi.$$

[1] Left-handed: which *unfolds*, or *unwinds*, in the anticlockwise direction.

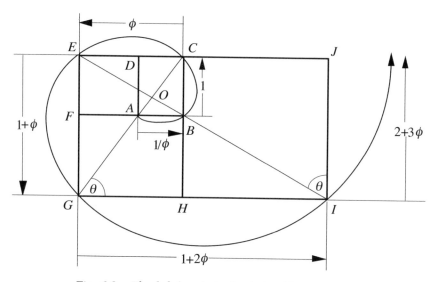

Fig. 6.2a. The left-handed whorled golden rectangle.

If the direction is reversed, we are in presence of a right-handed whorled rectangle, for which

$$\phi = \frac{1}{\phi - 1} = \frac{\phi - 1}{2 - \phi} = \frac{2 - \phi}{2\phi - 3} = \frac{2\phi - 3}{5 - 3\phi} = \dots \qquad (6.7)$$

Figure 6.2b highlights a remarkable property of the golden section: the addition of square *ADEF* to rectangle *ABCD*, whose vertical propor-

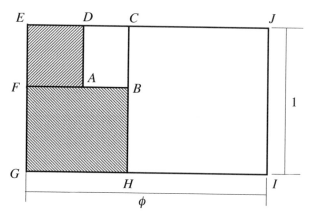

Fig. 6.2b. The golden rectangle's square gnomon.

tion[2] is the golden section, results in rectangle *BCEF*, whose horizontal proportion is also the golden section. Square *ADEF* is the gnomon of rectangle *ABCD*. Similarly, square *BFGH* is the gnomon of rectangle *BCEF*, and so on. That property obviously corresponds to the fact that all partial quotients are equal to 1 in continued fraction (6.3b).

THE FIBONACCI WHORL

Figure 6.3 shows a *left-handed whorl* with a square *seed* of unit size (colored gray). On its left, an adjacent square of unit size is drawn, followed by a square of size 2, then size 3, then sizes 5, 8, ... The process can continue indefinitely, generating at every step a new Fibonacci number.

It is important to observe that what looks like a logarithmic spiral envelope in that figure is not really logarithmic. Whereas the spiral of figure 6.2a possesses a pole 0, and generating diagonals CG and EF, upon which the successive vertices lie, there is no such thing in figure 6.2b. Here the spiral approaches a logarithmic spiral only following a large number of iterations. The whorl of figure 6.2a is *properly seeded*, whereas that of figure 6.3 is *improperly seeded*. The spiral envelope of a whorled figure is logarithmic only if the whorl is properly seeded.

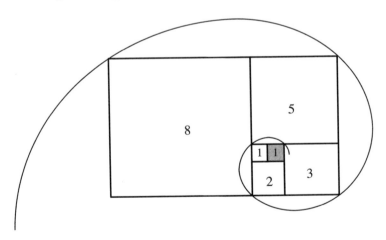

Fig. 6.3. The Fibonacci whorl.

[2] *Vertical proportion* means the ratio of the vertical to the horizontal sides. *Horizontal proportion* means the inverse of that ratio.

THE WHORLED GOLDEN TRIANGLE

There is good reason to believe that this triangle was
especially studied by the Pythagoreans, for it lies at the root
of many interesting geometrical constructions, such as the
regular pentagon and its mystical "pentalpha," and a whole
range of other curious figures beloved of the ancient
mathematicians: culminating in the regular pentagonal
dodecahedron, which symbolized the Universe itself, and
with which Euclidian geometry ends.
(D'Arcy Wentworth Thompson)[3]

Figure 6.4 illustrates a particularly beautiful example of gnomons
and self-similarity, showing successive generations of whorled isoceles
triangles. Two families of isoceles triangles can be observed in the
figure: a "sharp" family and a "blunt" family. Sharp triangle CBA has
a base angle of 72°, or $\pi/5$. The angle at vertex A is thus equal to 36°.
Sharp triangle DCB is similar to CBA, and the latter is obtained by
adding blunt triangle BAD to DCB. Triangle BAD is the gnomon of
triangle DCB, and its base angles are equal to 36°.

Close examination of the figure reveals that DCB was itself obtained
by adding to sharp triangle CED its own gnomon, namely, blunt triangle
CEB. This process can be endlessly repeated, with every sharp triangle
equal to the sum of a similar sharp triangle and its respective gnomon,
all of which gnomons are also similar. The ratio of the long side to the
base of a sharp triangle is none other than the magical golden section.
It is interesting to observe, en passant, that these triangles gave birth
to the two elementary figures (fig. 6.5) that Roger Penrose used to
irregularly tile the plane (plate 23). That triangle is what H. A. Naber
called the (unpronounceable) "Dreifachgleichenkelige Dreieck" in his
Das Theorem des Pythagoras (Haarlem, 1908).

THE WHORLED PENTAGON

Figure 6.6 shows the whorled pentagon, whose radius increases
by a factor equal to the Goden number for every 72° rotation, as each
golden triangle's base is equal to its predecessor's long side.

[3] On Growth and Form, p. 183.

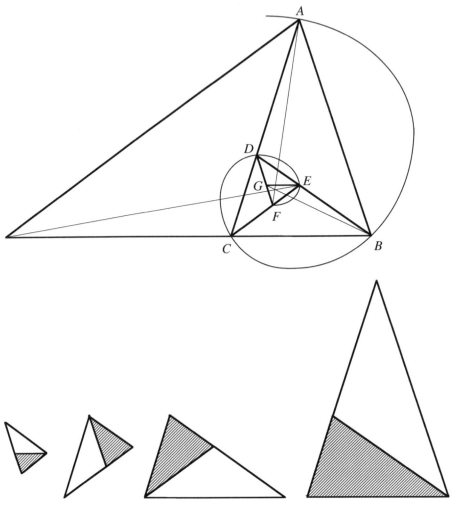

Fig. 6.4. The whorled golden triangle.

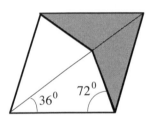

Fig. 6.5. Penrose's tiles.

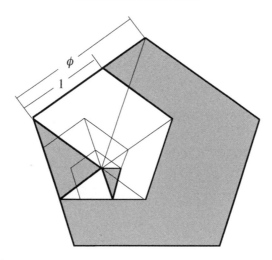

Fig. 6.6. The whorled pentagon, showing successive "divine triangles."

THE GOLDEN SECTION: FROM ANTIQUITY
TO THE RENAISSANCE

Let him not read my principles, who is not a mathematician.
(*Leonardo da Vinci*)

The golden ratio is present, or suggested, whenever a circle is divided into five or ten parts, a division that, since the dawn of humanity, was suggested to man by his five fingers, and the fivefold symmetry that characterizes innumerable varieties of flowers, seashells, and so on (fig. 6.7). Kepler was convinced that most flowers were five-petaled. Our ancestors may have therefore felt the presence of the golden ratio whenever they attempted to draw a regular convex pentagon, or a five-pointed star. Let us begin with the so-called Egyptian rectangle, shown in figure 6.8a. Its diagonal AC is equal to $\sqrt{1 + \phi} = \phi$, meaning that the sides AB, BC, CA of triangle ABC form a geometric progression of ratio $\sqrt{\phi}$. Moreover, triangles CDA and DEA are similar. Therefore,

$$\frac{AE}{AD} = \frac{AD}{AC} \therefore AE = \frac{(AD)^2}{AC} = 1 \therefore \frac{AE}{AC} = \phi.$$

In other words, the rectangle's diagonal is divided according to the middle and extreme ratios at point E.

Fig. 6.7. Fivefold symmetry exhibited by the sand dollar.

It was reported that the Greek historian Herodotus learned from the Egyptian priests that the square of the great pyramid's height is equal to the area of its triangular lateral side (fig. 6.8b), meaning that the ratio of the pyramid's height h to $b/2$, half the side of its square base, is precisely equal to the golden ratio. It is to this day a matter of controversy whether the Egyptians knew the golden ratio as such, or just stumbled upon it, as many detractors of Egyptian genius affirm. Be that as it may, the pyramid's original height h, according to the most reliable estimates, measured 280 royal cubits, and its base b measured 440 cubits.[4] The ratio $h/(b/2)$ is therefore equal to $14/11 = 1.27272\ldots$. Compare that number to $\sqrt{\phi} \approx 1.27202$. The difference is a mere 0.06%. That is probably why the triangle of figure 6.8a is

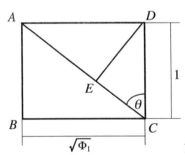

Fig. 6.8a. The "Egyptian" triangle.

<hr />

[4] A royal cubit is approximately equal to 52.5 cm, the measure of a forearm. A royal cubit was divided into seven palms of 7.5 cm, each consisting of four fingers of 1.875 cm. The word *cubit* comes from the Latin *cubitum*, meaning elbow. The French sometimes use a similar archaic measure called the *coudée*.

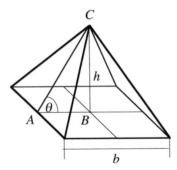

Fig. 6.8b. The Egyptian pyramid.

called the Egyptian triangle! According to R. J. Gillings, the slope θ, or *seked*, of the pyramid's triangular side is 51° 52'. Compare that number to $\tan^{-1} \sqrt{\phi} = 51°\ 49'\ 38''$.

Moving on to the Greeks, the Parthenon, which was dedicated in 438 B.C. by architects Actinus and Callicrates in collaboration with sculptor Phidias, is regarded by some authors as the first embodiment of the golden ratio in architecture. The temple is allegedly inscribed, as in figure 6.9a, in a golden rectangle, and served as a model for subsequent Greek constructions.

It is not until Euclid, however, that the golden ratio's mathematical properties were studied. In the *Elements* (308 B.C.) the Greek mathematician merely regarded that number as an interesting irrational number, in connection with the middle and extreme ratios. Its occurrence in regular pentagons and decagons was duly observed, as well as in the dodecahedron (a regular polyhedron whose twelve faces are regular pentagons). It is indeed exemplary that the great Euclid, contrary to generations of mystics who followed, would soberly treat that number for what it is, without attaching to it other than its factual properties.

The Middle Ages witnessed little progress of the sciences in Europe, despite the innumerable advances made by the Arabs. One of the outstanding figures of that period was Leonardo of Pisa, the familiar Fibonacci, author of the *Liber Abaci* (1202) and discoverer of the famous sequence that bears his name. Another notable figure of the middle ages was Johannes Campanus of Novare, whose immense merit was the translation of Euclid's *Elements* into Latin, from an Arabic version of the monumental treatise (ca. 1260). In a comment accompanying the translation, reference is made to the middle and extreme ratio, *proportionem habentem medium duoque extrema.*

Over the ages, the five-pointed star, or *pentagram* or *pentalpha*, became the object of superstitious beliefs, as reflected by the Gothic

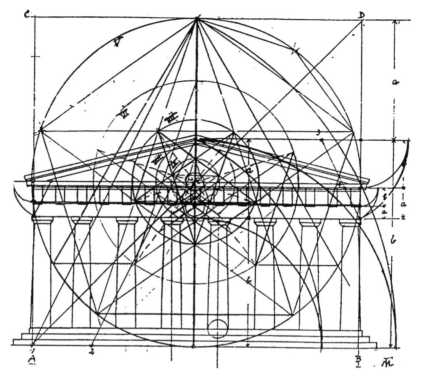

Fig. 6.9a. The Parthenon, drowned under a wildly imaginative array of lines, circles, and pentalphas, wherein one is supposed to recognize the presence of the golden ratio in the temple's proportions. From Marguerite Neveux, *Le Nombre d'or: radiographie d'un mythe* (Paris: editions du Seuil, 1995).

text of figure 9b. The five pointed-star also evolved into an arabesque motif (fig. 6.9c).

The golden era of the golden number was the Italian Renaissance. The expression *divine proportion* was coined by the great mathematicians Luca Pacioli in his book *De divina proportione*, written in 1509. Born in Borgo San Sepulcro in Tuscany ca. 1445, he studied mathematics, notably under the guidance of Piero della Francesca, author of *De quinque corporis regularibus* (The Five Regular Bodies). He then studied philosophy and theology, before becoming a Franciscan monk. In 1496 he was offered the Milan mathematics chair by Duke Ludovic Sforza. He was forced to leave for Florence in 1498 upon his protector's death.

The pentagram, a five-pointed star drawn with one stroke of the pen: this sign belongs, as do many others depicted here, to the most primitive of mankind, and is certainly much older than written characters. Signs of this kind are quite the most ancient human documents we possess. The pentagram has had several different significations at different times in the history of man. The Pythagoreans called it the pentalpha, and the Celtic priests the witch's foot. It is also Solomon's seal, known in the Middle Ages as the goblin's cross.

Fig. 6.9b. The star pentagram. From Rudolf Koch, *The Book of Signs* (New York: Dover, 1930).

Fig. 6.9c. The five-pointed star is an arabesque pentagram.

Fig. 6.10a. Portrait of Luca Pacioli by Iacopo de Barberi.
Museo Capodimonte, Naples.

In 1514, the year of his own death, he was called in Rome by Pope
Leo X. *De divina proportione* is essentially inspired by Euclid's *Elements*
and devotes forty-seven chapters to the Platonic solids. A famous portrait
of Pacioli by Iacopo de Barberi (1495) may be admired at the Naples
museum (fig. 6.10a). It represents the mathematician with his left hand
showing his textbook, while his right hand draws a triangle inscribed
in a circle, as it appeared in Euclid's *Elements*. On the table lies a
dodecahedron, and from the ceiling hangs a transparent twenty-six-
sided polyhedron.

In the eyes of Pacioli, beside being the essence of the pentagon
and decagon, as well as the dodecahedron, which according to Plato
represented the universe, the golden proportion was unique and myste-
rious, in the image of God. It defined an immutable relationship between
one another, of a (holy) trinity of magnitudes. Its irrationality also
conveyed to the section a divine character, descended from the heavens.
Pacioli exerted considerable influence on the thinkers and artists of his

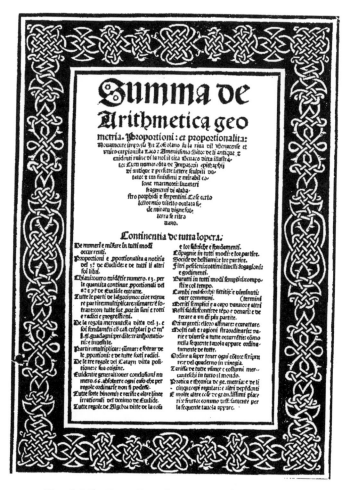

Fig. 6.10b. Luca Pacioli's *Summa de arithmetica*.

time. Leonardo da Vinci's encounter with Luca Pacioli was a turning point in the painter's life, as he discovered at the age of forty the beauty and enjoyment of mathematics. Leonardo became so fascinated by the subject that he studied in great detail the *Summa de arithmetica, geometria proportioni e proportionalita,* authored by Pacioli in 1487 (fig. 6.10b) and copied the fastidious *genealogical tree of proportions,* which contained no less than forty categories, organized in classes and subclasses. It is believed that it was Leonardo who coined the expression *sectia aurea,* or golden section. Leonardo also studied Euclid's *Elements* and unfortunately wasted considerable time attempting to square the

.C V.

DVODECEDRON PLANV
VACVVS.

XXVIII

Fig. 6.11. The dodecahedron drawn by Leonardo to illustrate Pacioli's
De divina proportione.

circle. Upon publishing his *De divina proportione,* Pacioli balked at
the difficult task of precisely drawing the Platonic solids. Instead,
he built wooden models with sticks of *vile material.* It is believed
that Leonardo illustrated the book at Pacioli's request, leaving to
posterity sixty marvelous polyhedra (fig. 6.11). It is also reported
that Albrecht Dürer visited Pacioli in Bologna to learn the "secret
perspective" from the great master.

Much has been written in modern times about the role played in
art by the golden section, ranging from the analysis of ancient works
in light of the section such as that of van der Weyden's *Descent from
the Cross* (Prado Museum, Madrid) by the French architect Le Corbusier,
to the deliberate use of the section in painting, such as Seurat's *Parade
de Cirque,* and the extensive use of the *modulor,* a contraction of *module
d'or,* by Le Corbusier.

MARGINALIA

The Sneezewort[5]

Imagine that your gardener plants one morning a mythical triadic tree in your garden. From every branch of the tree, three new branches sprout every year. The gardener carefully attaches the branches to the quickset hedge following an immutable pattern, one in the center, and one on either side. If left unpruned, the number of extremities at the end of the first, second, third year, and so on would be 1, 3, 9, and so on. Your gardener, who is known for his numerous eccentricities, decides, on even years, to cut off every branch whose direction veers to the left of that of its predecessor, and on odd years every branch that veers to its predecessor's right. The pruning program is shown in figure 6.12a, and the resulting tree, at the end of the seventh year,

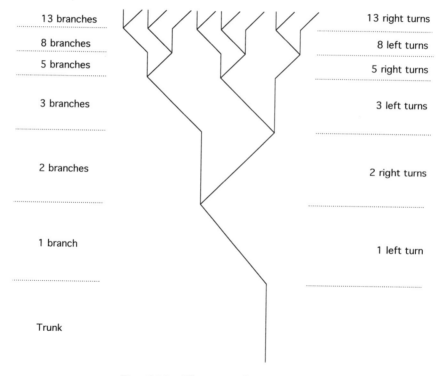

13 branches		13 right turns
8 branches		8 left turns
5 branches		5 right turns
3 branches		3 left turns
2 branches		2 right turns
1 branch		1 left turn
Trunk		

Fig. 6.12a. The pruned triadic tree.

[5] I first discovered this strange plant, whose scientific name is *Achillea ptarmica*, in Professor H. E. Huntley's beautiful little book *The Divine Proportion* (New York: Dover, 1970).

Fig. 6.12b. An imaginary sneezewort.

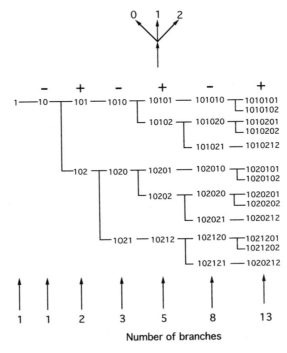

Fig. 6.12c. The pruned triadic number tree.

looks somewhat like the imaginary *sneezewort* of figure 6.12b, where the missing branches are replaced by leaves. The number of branches belonging to each generation are none other than the Fibonacci numbers of the first order, namely, 1, 1, 2, 3, 5, 8, 13, ... figure 6.12c shows how to achieve the same result using triadic numbers. In the columns marked (+), every allowable integer *larger* than the number's rightmost integer is added to its right, and in the columns marked (−), every allowable integer *smaller* than the rightmost integer is added to its right.

Phyllotaxis, the study of leaf growth, abounds with occurrences of the golden number. In the introduction, we mentioned the divergence in the aloe plant, the sunflower's florets, and the scales of the pinecone. The sneezewort obviously offers another striking illustration. Professor

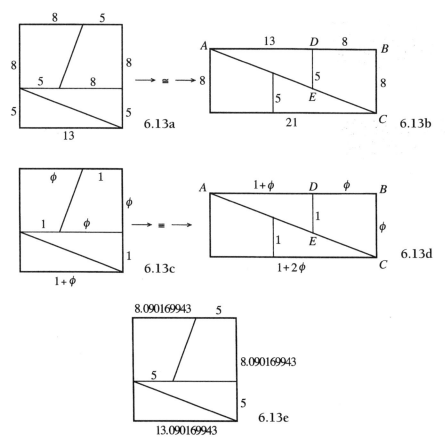

Fig. 6.13. A golden trick.

Huntley's book presents many very interesting properties of the golden number. He draws a parallel between the reproducing patterns of the bee, as well as the histogram of a hydrogen atom's energy states, with the tree in figure 6.12a. Another classical example is offered by the multiple reflections between three semireflecting glass plates.

A Golden Trick

Figure 6.13 offers a deceptive geometric construction, in which the rectangle of figure 6.13b is obtained by what seems to be a re-arrangement of the pieces of figure 6.13a. In so doing, the figure's area is seen to shrink from $13^2 = 169$ units to $8 \times 21 = 168$ units! Where did the one unit of area disappear? The fallacy lies in that, despite appearances, AEC is not a straight line in 6.13b. Had it been a straight line, we would have had $DE/BC = AD/AB$. That is not the case, as $5/8 \neq 13/21$. Close, but not quite. The answer is given by figures 6.13c and 6.13d, in which $(1 + 2\phi)/(1 + \phi) = \phi$. Figure 6.13e is a redrawing of figure 6.13c to the same scale as that of figure 6.13a.

The Golden Knot

Figure 6.14 shows how the mere act of tying a strip of paper into an ordinary knot generates a perfect pentagon. It is also interesting to observe that upon joining extremities A and B, a Moebius strip is obtained, as the knot achieves a full 180° twist of the paper strip.

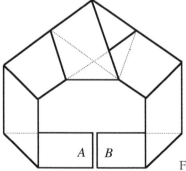

Fig. 6.14. The golden knot.

The Silver Number

Tales of a neglected number
(*Ian Stewart*)[1]

Architect Richard Padovan discovered an integer sequence that, were it not for its interesting geometric metaphor and the striking parallels that can be drawn with the golden number, might have been dismissed as rather trivial, hence the unkind name given to it, the "plastic" sequence. Trivial it is not, as we shall see. The sequence is defined by recursion:

$$P_{i+3} = P_i + P_{i+1} \tag{7.1}$$

with

$$P_0 = P_1 = 0, \qquad P_2 = 1. \tag{7.2}$$

The resulting sequence thus begins as follows, and may be extended for positive as well as negative values of i:

i	-3	-2	-1	0	1	2	3	4	5	6	7	8	9	10	11	12	13	14	15	16	17	18	19
P_i	1	-1	1	0	0	1	0	1	1	1	2	2	3	4	5	7	9	12	16	21	28	37	49

Any set of three consecutive integers from the sequence may serve as initial conditions. We have found the above choice to be consistent with the analysis which follows. Let us first consider equation

$$p^3 - p - 1 = 0, \tag{7.3a}$$

which is reminiscent of equation

$$\phi^2 - \phi - 1 = 0, \tag{7.3b}$$

whose solution is the golden number ϕ.

[1] "Tales of a Neglected Number," *Scientific American* (June 1996), pp. 92–93.

An approximate value of p satisfying equations (7.3) can be computed to any desired number of decimals, by iterating the expression

$$\sqrt[3]{1 + \sqrt[3]{1 + \sqrt[3]{1 + \cdots \sqrt[3]{1 + s}}}} \to p, \qquad (7.4)$$

where s is any arbitrary "seed" value. Iteration can be performed on a handheld calculator, until two consecutive values are identical. With ten significant decimal digits, we find

$$p \approx 1.324717957. \qquad (7.5)$$

This has been referred to in recent literature as the *plastic number*. Notwithstanding the number's detractors, it has no lesser claim to nobility than its illustrious golden ancestor, and we shall refer to it as the *silver number*. The Greek letter ϕ had been chosen to honor Phidias, the Pantheon's sculptor. It is appropriate that p, the Latin initial of architect Padovan, should be retained to represent the silver number.[2]

Substituting $p^3 = p + 1$ in the sequence

$$\cdots \frac{1}{p^2}, \frac{1}{p}, 1, p, p^2, p^3, p^4, \ldots,$$

we obtain

$$\ldots (-p^2 + p + 1), (p^2 - 1), 1, p, p^2, (p + 1), (p^2 + p), \ldots.$$

Comparing the coefficients of $1, p, p^2$ in the terms of the above sequence to the integers of the Padovan sequence, we find by inspection

$$p^i = P_i p^2 + P_{i+1} p + P_{i-1}, \qquad (7.6a)$$

a statement that can be proven by induction over i. The reader can compare that statement to

$$\phi^i = F_i \phi + F_{i-1}, \qquad (7.6b)$$

which expresses the ith power of the golden number in terms of the Fibonacci sequence numbers. It follows from equations (7.6) that p is a real solution of equation

$$x^i - P_i x^2 - P_{i+1} x - P_{i-1} = 0 \qquad (7.7)$$

for all values of i. For example,

[2] If one is inclined to believe in signs, he or she will note that the kinship between the two numbers is more surprising than initially meets the eye. As Ian Stewart points out, the Italian town of Padova is roughly 100 miles from Pisa, the home of Leonardo, who was otherwise known as Fibonacci!

$$x^5 - x^2 - x - 1 = 0, \qquad x^7 - 2x^2 - 2x - 1 = 0. \quad (7.8)$$

As in the Fibonacci sequence, the ratio between two successive Padovan numbers converges, though more slowly, to p. For large values of i, we get

$$p \approx \frac{P_{i+1}}{P_i}. \qquad (7.9a)$$

Using equation (7.6), the reader can verify the following statement

$$(3p^2 - 1)P_i - p^i = 2P_i p^2 - P_{i+1}p - P_{i+2}.$$

For large values of i, applying (7.9a), we have $P_i p^2 \approx P_{i+1}p \approx P_{i+2}$, hence $2P_i p^2 - P_{i+1}p - P_{i+2} \approx 0$, and we get

$$P_i \approx \frac{p^i}{(3p^2 - 1)} \quad \text{to be compared to} \quad F_i \approx \frac{\phi^i}{(2\phi - 1)}, \quad (7.9b)$$

and we also obtain

$$p^i \approx 2P_{i+1}p + P_{i-1} \quad \text{to be compared to} \quad \phi^i \approx 2F_{i+1} - F_i. \quad (7.9c)$$

For example,

$$(7.9a) \rightarrow \frac{P_{19}}{P_{18}} = \frac{49}{37} \approx 1.324324324 \approx p,$$

$$(7.9b) \rightarrow \frac{p^{19}}{3p^2 - 1} \approx \frac{209.0956721}{4.26432997} \approx 49.03365208 \approx P_{19},$$

$$(7.9c) \rightarrow 2P_{20}p + P_{18} = 209.2133344 \approx p^{19}.$$

FROM NUMBER TO GEOMETRY

Figure 7.1a consists of a right-handed whorl of equilateral triangles, which begins with a seed consisting of three identical unit triangles. The triangles that follow are of sizes 2, 2, 3, 4, 5, ... Triangle $i + 3$ is equal to the sum of triangles i and $i + 1$. The triangles therefore follow one another according to the Padovan sequence, beginning with P_4. For comparison purposes, figure 7.1b shows the left-handed Fibo-

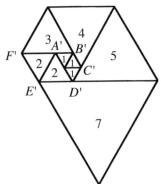

Fig. 7.1a. The Padovan triangular whorl.

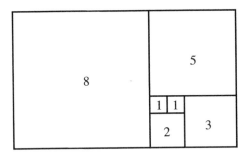

Fig. 7.1b. The Fibonacci rectangular whorl.

nacci rectangular whorl, in which the reader may recognize the consecutive Fibonacci numbers 1, 1, 2, 3, 5, 8, ... Under the heading "Tales of a Neglected Number," Ian Stewart, to whom I am grateful for allowing me to discover the Padovan sequence, inscribes a spiral *inside* the Padovan whorl, recognizing that it only approaches a true logarithmic spiral. In what follows, we shall construct a different kind of whorl, which is *circumscribed* by a bona fide logarithmic spiral, following a construction akin to that of figure 7.1a, though distinct from it.

THE SILVER PENTAGON

We now introduce the truncated parallelogram $ABCDE$ of figure 7.2a, which we shall call the *silver pentagon*. Sides AB, BC, CD, DE, EA are respectively equal to 1, p, p^2, p^3, p^4. Angle AED is equal to $60°$. By extending sides AB and CD until they intersect at Q, we can verify that $DE = AB + BQ$, that is, $p^3 = p + 1$, and $AE = QC + CD$, that is, $p^4 = p^2 + p$. With the lengths chosen for the truncated parallelogram,

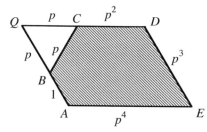

Fig. 7.2a. The silver pentagon.

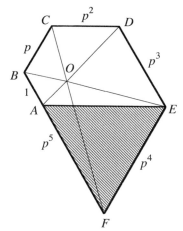

Fig. 7.2b. The pentagon's gnomon.

the construction is therefore perfectly legitimate. If we now add equilateral triangle AEF as in figure 7.2b, we obtain the truncated parallelogram $BCDEF$, whose sides BC, CD, DE, EF, FB are respectively equal to p, p^2, p^3, p^4, p^5. It is a silver pentagon, whose size is that of $ABCDE$ multiplied by p. Equilateral triangle AEF is therefore the *gnomon* of the silver pentagon. Figure 7.3 shows a succession of figures in which the addition of an equilateral triangle to any silver pentagon results in a new silver pentagon enlarged by a factor p.

THE SILVER SPIRAL

Figure 7.4a shows a right-handed whorl around pole O, starting with *seed ABCDE*. Observe that diagonals AD, BE, CF intersect at pole O. The whorl is properly seeded, and the resulting spiral envelope is logarithmic. Its radius grows by factor p with every 60° rotation. Its polar coordinate equation is therefore $r = r_0 e^{\lambda\vartheta}$ with flare coefficient

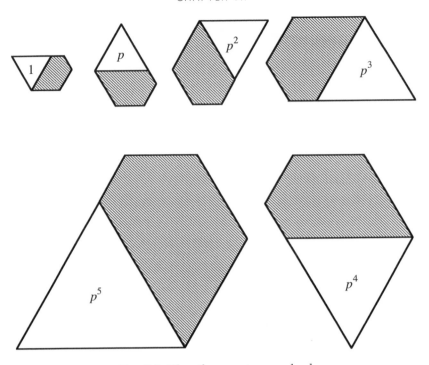

Fig. 7.3. The silver pentagon whorl.

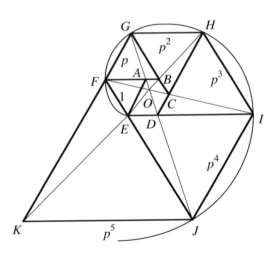

Fig. 7.4a. The silver spiral.

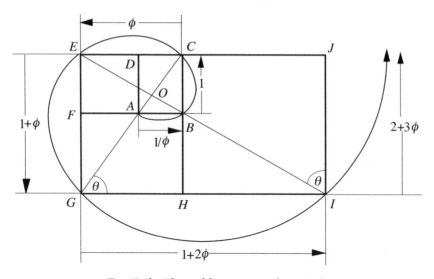

Fig. 7.4b. The golden rectangular spiral.

$$\lambda = \frac{3\ln(1.324718)}{\pi} = 0.8834145. \qquad (7.10)$$

For comparison purposes, figure 7.4b shows the left-handed golden rectangular whorl. The radius of its logarithmic spiral envelope grows by a factor ϕ with every 90° rotation. Its flare coefficient

$$\lambda = \frac{2\ln(1.618034)}{\pi} = 0.30635. \qquad (7.11)$$

If we compare figures 7.1a and 7.4a, we observe that though pentagons $ABCDE$ and $A'B'C'D'E'$ look alike, they are obviously not similar. The addition of equilateral triangle $A'E'F'$ to $A'B'C'D'E'$ generates pentagon $B'C'D'E'F'$, which is *not* similar to $A'B'C'D'E'$. That pentagon does not have the equilateral triangle for a gnomon. As the whorl proceeds, more and more equilateral triangles are added, and the truncated pentagons of figure 7.1a more and more closely resemble the silver pentagon, as the ratio P_i/P_{i-1} tends to p when i is sufficiently large. Similarly, if we compare figures 7.1b and 7.4b, we find that only the latter is properly seeded and can be circumscribed by a logarithmic spiral.

THE WINKLE

The geometric figure shown in figure 7.5 was obtained by adding equilateral triangles to one another in succession, with each triangle half the size of its predecessor. *If the process is extended to infinity*, and only in that case, we shall call the resulting figure a *triangular winkle*,[3] by virtue of its resemblance with the mollusk so named, which inhabits a spiral shell from which it is not easy to "winkle out"! Figure 7.6 shows how the addition of an equilateral triangle to the winkle results in a new winkle, every side of which is twice that of its predecessor. Had not the process of figure 7.5 been infinite, the figure resulting from the addition of an equilateral triangle to the truncated winkle would not have been geometrically similar to the mutilated mollusk. Every addition of an equilateral triangle would have resulted in a new figure, not geometrically similar to any of its predecessors.

The character of self-similarity, or gnomonicity, of the winkle allows

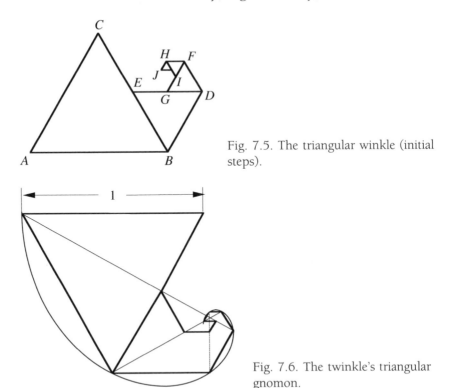

Fig. 7.5. The triangular winkle (initial steps).

Fig. 7.6. The twinkle's triangular gnomon.

[3] The facetious reader may wish to refer to it as a *twinkle*.

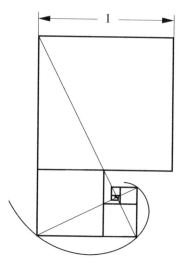

Fig. 7.7. The squinkle's gnomon.

it to be circumscribed by a logarithmic spiral, as shown in figure 7.6, whose radius doubles every 60°. Its flare is thus $\lambda = 3/\pi \ln 2 \approx 0.661906801$. Figure 7.7 illustrates how a square winkle[4] can be obtained following a similar process. The spiral's flare in this case is $\lambda = 2/\pi \ln 2 \approx 0.4412712$. Although their construction process is infinite, the triangular and square winkles have finite perimeters. The triangular winkle's perimeter is $2(1 + \frac{1}{2} + \frac{1}{4} + \cdots) = 4$, and that of the square winkle is 6. Similarly, their areas are convergent series, hence also finite.

Notwithstanding the finite character of both winkles' perimeters and areas, we shall not refer to them as *finitary* polygons and reserve that epithet to those polygons that consist of a finite number of finite sides. Hence the following remarkable statements:

> The square is the golden rectangle's gnomon, and the equilateral triangle the silver pentagon's gnomon. There is (in all likelihood) no finitary polygon whose gnomon is a regular polygon other than the golden rectangle and the silver pentagon.

MARGINALIA

Golomb's Rep-Tiles

Certain families of figures provide insight into the study of gnomons. Among these, we find *rep-tiles*, a word coined by the American mathe-

[4] Or *squinkle*.

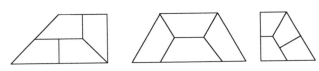

Fig. 7.8. Golomb's rep-4 figures.

matician Solomon Golomb to encompass these intriguing forms, which were published by Martin Gardner in his *Scientific American* column, then compiled in his collection *The Unexpected Hanging*,[5] along with so many other fascinating mind-expanding essays.

Figure 7.8 shows three trapezoids, each consisting of four congruent trapezoids, all of which are geometrically similar to the larger figure, and one-fourth as large. These trapezoids are said to be rep-4, or of replicating order 4, meaning that four identical figures are required, following specified assembly rules, to generate a larger figure that is similar to the smaller one. The process can be repeated any number of times, and the elementary trapezoids can thus be made to regularly tile a floor, hence the name given by Golomb to the form.

Another interesting property discovered by Golomb is that the $1 \times \sqrt{k}$ parallelogram is rep-k, as illustrated by figure 7.9. It takes k identical $1 \times \sqrt{k}$ polygons to generate a similar polygon. In terms of

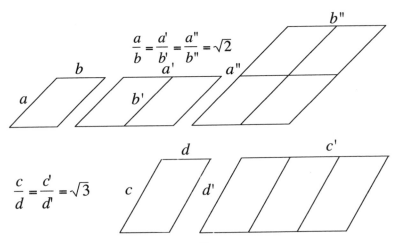

Fig. 7.9. The 1 by $\sqrt{k}$ polygons are rep-k.

[5] *The Unexpected Hanging* (New York: Simon and Schuster, 1969).

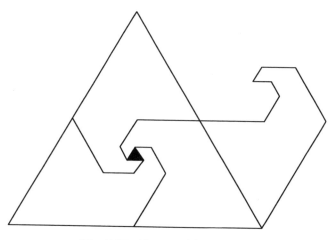

Fig. 7.10. The twinkle is rep-4.

gnomons, Golomb's observation is equivalent to the statement

$$\Phi_m = \sqrt{k} \quad \text{or} \quad m = \Omega_m - \frac{1}{\Phi_m} = \frac{k-1}{\sqrt{k}}.$$

Given a seed parallelogram of sides 1 and $\sqrt{k}$, that is, of proportion $\sqrt{k}$, its gnomon shares side $\sqrt{k}$ with the seed, and its other side measures

$$\Phi_m \sqrt{k} = \frac{(k-1)\sqrt{k}}{\sqrt{k}} = k - 1.$$

The gnomon therefore consists of $k - 1$ parallelograms, all of which are congruent to the seed. The sum (seed + gnomon) is equal to the seed enlarged k times. Another of Golomb's discoveries, reexpressed here in terms of gnomons, is that the only two *homognomonic* figures (congruent to their gnomon) are the parallelogram of proportion $\sqrt{2}$, and the isoceles right triangle.

Many years ago, the prolific Golomb also discovered the form shown in figure 7.10, which is none other than the triangular winkle, or twinkle. It is rep-4, as the three inner twinkles interlock perfectly, provided that the innermost little triangular gap vanishes, as the construction is carried to infinity.

Fig. 7.11. Girolamo Cardano (Jerome Cardan).

A Commedia dell'Arte

In his momentous *Ars Magna*, published in 1545, the Italian mathematician Girolamo Cardano (1501–1576; fig. 7.11) offered the following solution to the cubic equation $x^3 + ax = b$:

$$x = \sqrt[3]{\sqrt{\left(\frac{a}{3}\right)^3 + \left(\frac{b}{2}\right)^2} + \frac{b}{2}} - \sqrt[3]{\sqrt{\left(\frac{a}{3}\right)^3 + \left(\frac{b}{2}\right)^2} - \frac{b}{2}}. \qquad (7.12)$$

Equation $p^3 - p - 1 = 0$ can be solved by substituting $a = -1$, $b = 1$ in this equation, yielding

$$p = \sqrt[3]{\sqrt{\left(\frac{-1}{3}\right)^3 + \left(\frac{1}{2}\right)^2} + \frac{1}{2}} - \sqrt[3]{\sqrt{\left(\frac{-1}{3}\right)^3 + \left(\frac{1}{2}\right)^2} - \frac{1}{2}}$$

$$= \sqrt[3]{\sqrt{\frac{23}{108}} + \frac{1}{2}} - \sqrt[3]{\sqrt{\frac{23}{108}} - \frac{1}{2}} \quad (7.13)$$

$$\approx \sqrt[3]{0.461479103 + 0.5} - \sqrt[3]{0.461479103 - 0.5}$$

$$\approx 0.986991206 + 0.337726751$$

$$\approx 1.324717957,$$

which is identical, within ten decimal places, to the value calculated through successive iterations of the expression

$$\sqrt[3]{1 + \sqrt[3]{1 + \sqrt[3]{1 + \cdots \sqrt[3]{1 + s}}}} \to p. \quad (7.14)$$

As Carl B. Boyer notes,[6] the year 1545 marks the beginning of the modern period of mathematics. For only one short year, until the publication of *Ars Magna*, the definitive book on algebra had heretofore been Michael Stifel's *Arithmetica Integra*, published in 1544, following the *Coss* (1525) and the *Rechnung* (1527), authored respectively by Christoff Rudoff and Peter Apian. Stifel, Coss, and Apian were the most prominent members of the German school of mathematics, which was particularly prolific during that era. It is also worth noting that in the year 1543, Nicholas Copernicus published his *De revolutionibus orbium coelestum*, and Andreas Vesalius his *De Humani Corporis Fabrica* (On the Fabric of the Human Body).

While it is true that Cardano gave credit for the solution of the cubic equation to his compatriot Niccolo Fontana (1500–1557), he also stole his thunder. As it turned out, Cardano had promised Fontana, upon the latter's revelation of the solution, that he would refrain from making it public, as Fontana was furbishing his own treatise on algebra, with the equation's solution as its masterpiece. But that was not the only episode of a true sixteenth-century commedia dell'arte, whose central protagonists were Cardano and Fontana, nicknamed Tartaglia, or stutterer. The impediment suffered by the unfortunate Tartaglia had

[6] Carl B. Boyer and Uta C. Merzbach, *A History of Mathematics* (New York: John Wiley & Sons, 1989), p. 282.

been caused by a sabre blow received in 1512, upon the fall of the city of Brescia to the French. Tartaglia had the bad habit of ascribing other people's inventions to his name. He lifted his Archimedean translation from an author named Moerbeke, without giving him credit, and it is quite probable that he plagiarized the law of the inclined plane discovered by Jordanius Nemorarius. It is equally probable that the solution of the cubic equation was inspired to Tartaglia by the obscure Scipione del Ferro, professor of mathematics at the very respectable University of Bologna, who had disclosed his secret to his student Fiore (Floridus in Latin). Following a mathematical contest between Fiore and Tartaglia, easily won by the latter, Tartaglia was invited by Cardano to his home, lured by the prospect of being introduced to a patron, whose support he badly needed. In his *Ars Magna*, Cardano also gave the solution of the quartic equation, which, according to him, was invented by Luigi Ferrari "at his request." Initially a physician, Cardano, a native of Pavia, was an adventurer and a gambler. As an astrologer, he went as far as casting the Christ's horoscope! Although a miscreant and an illegitimate child, he obtained a pension from the Pope. His family's saga would fill a volume, with one son a knave, and the other the assassin of his own wife. Nonetheless, Cardano, a faithful disciple of Al Khawarizmy, remains the best algebraist of his time.

Repeated Radicals

So far, we have examined two kinds of iterative forms capable of generating numbers, namely, the positional number representation and the continued fraction. In this book up to this point we have discovered a new iterative form consisting of *repeated radicals*, examples of which are

$$\sqrt{1 + \sqrt{1 + \sqrt{1 + \sqrt{1 + \cdots}}}} \to \phi \qquad (7.15)$$

and

$$\sqrt[3]{1 + \sqrt[3]{1 + \sqrt[3]{1 + \sqrt[3]{1 + \cdots}}}} \to p. \qquad (7.16)$$

These expressions can be regarded as particular cases of the general form

$$\sqrt[n]{b + a\sqrt[n]{b + a\sqrt[n]{b + a\sqrt[n]{\ldots}}}} \qquad (7.17)$$

Assuming that the above expression converges to some value x, we can write

$$x = \sqrt[n]{b + a\sqrt[n]{b + a\sqrt[n]{b + a\sqrt[n]{\ldots}}}}. \qquad (7.18)$$

The metaphor known as Hilbert's hotel has it that if you own a hotel with an infinite number of rooms, all of which are full, you can always accommodate a newcomer by asking the tenant of room 1 to move to room 2, that of room 2 to room 3, and so on, without any tenant ever falling off the roof. We may resort to that metaphor in the present case and decide that since $x = \sqrt[n]{\ldots}$, we are authorized to move him to the next room, and write $x = \sqrt[n]{b + ax}$, which yields

$$x^n - ax - b = 0. \qquad (7.19)$$

Equations (7.15) and (7.16) obviously correspond to $a = b = 1$.

In chapter III we saw that to equation

$$x^n - F_{m,n}x - F_{m,n-1} = 0 \quad \text{and in particular} \quad x^2 - mx - 1 = 0 \quad (7.20)$$

corresponds the solution $x = \Phi_m = (m + \sqrt{m^2 + 4})/2$ for all values of n. Hence,

$$\Phi_m = \sqrt[n]{F_{m,n-1} + F_{m,n}\Phi_m}, \quad \text{for all } n \neq 0, 1, \qquad (7.21)$$

which, following Hilbert's prescription, becomes

$$\Phi_m = \sqrt[n]{F_{m,n-1} + F_{m,n}\sqrt[n]{F_{m,n-1} + F_{m,n}\sqrt[n]{F_{m,n-1} + F_{m,n}\sqrt[n]{\ldots}}}}. \qquad (7.22)$$

For example, remembering that $F_{1,1} = 1$, $F_{1,2} = 1$, $F_{1,3} = 2$, $F_{1,4} = 3$, $F_{1,5} = 5, \ldots$, we get

$$\phi = \Phi_1 = \sqrt{1 + \sqrt{1 + \sqrt{1 + \sqrt{1 \ldots}}}}$$
$$= \sqrt[3]{1 + 2\sqrt[3]{1 + 2\sqrt[3]{1 + 2\sqrt[3]{\ldots}}}}$$
$$= \sqrt[4]{2 + 3\sqrt[4]{2 + 3\sqrt[4]{2 + 3\sqrt[4]{\ldots}}}}$$
$$= \sqrt[5]{3 + 5\sqrt[5]{3 + 5\sqrt[5]{3 + 5\sqrt[5]{\ldots}}}},$$
$$\ldots \qquad (7.23)$$

For $m = 2$, $\Phi_2 = 1 + \sqrt{2}$, equation (21) becomes

$$\Phi_2^n - F_{2,n-1}\Phi_2 - F_{2,n} = 0.$$

The Fibonacci sequence of order $m = 2$ is 0, 1, 2, 5, 12, 29, ... This allows us to write

$$1 + \sqrt{2} = \sqrt{1 + 2\sqrt{1 + 2\sqrt{1 + 2\sqrt{\ldots}}}}$$
$$= \sqrt[3]{2 + 5\sqrt[3]{2 + 5\sqrt[3]{2 + 5\sqrt[3]{\ldots}}}} \qquad (7.24)$$

An amusing, though perhaps not very useful, result is the following. Given that

$$\sinh x = \frac{e^x - e^{-x}}{2} = \frac{1}{2}\left(e^x - \frac{1}{e^x}\right),$$

we can write $m = 2 \sinh x$, hence, for positive x values,

$$e^x = \sqrt{1 + 2\sinh x \sqrt{1 + 2\sinh x \sqrt{1 + 2\sinh x\sqrt{\ldots}}}}. \qquad (7.25)$$

We can also write, for what it's worth,

$$e^x = 2\sinh x + \cfrac{1}{2\sinh x + \cfrac{1}{2\sinh x + \cfrac{1}{2\sinh x + \ldots}}}. \qquad (7.26)$$

It is not possible to close a discussion of repeated radicals without mentioning the famous formula offered in 1593 by the French mathematicians François Viète in his *Variorum de rebus mathematicis responsorum liber VIII*:

$$\frac{2}{\pi} = \sqrt{\frac{1}{2}} \cdot \sqrt{\frac{1}{2} + \frac{1}{2}\sqrt{\frac{1}{2}}} \cdot \sqrt{\frac{1}{2} + \frac{1}{2}\sqrt{\frac{1}{2} + \frac{1}{2}\sqrt{\frac{1}{2}}}} \cdots \qquad (7.27)$$

as well as the following beautiful formula discovered by Ramanujan:

$$\sqrt{1 + 2\sqrt{1 + 3\sqrt{1 + 4\sqrt{1 + \cdots}}}} = 3. \qquad (7.28)$$

If you are now asked to produce an *integer* x that satisfies the statement

$$\sqrt{x + \sqrt{x + \sqrt{x + \cdots}}} = x, \qquad (7.29)$$

you should have no difficulty coming up with the answer.[7]

[7] $x = 2$.

Spirals

> The spiral is a spiritualized circle. In the spiral form, the
> circle, uncoiled, unwound, has ceased to be vicious;
> it has been set free.
> *(Vladimir Nabokov, Speak Memory)*

W̲hat is a spiral, and why, aside from Bernoulli's fascination
with its "miraculous" character, is it relevant to the study of iterative
processes and gnomons? This chapter is an attempt to answer that
question. Here we address the problem of expressing a spiral's equation
in terms of Cartesian and polar coordinates. We then generalize our
study to that of dignomonic spirals and close the chapter with a finite-
difference approach to the study of oscillations in linear systems, illustrated by an analysis of the simple pendulum and the electrical resistance-inductance-capacitance (RLC) circuit. In so doing, we shall follow
a three-step approach from the simple to the more complex, instead of
going from the general to the particular, as any good "Cartesian" would.

THE ROTATION MATRIX

Consider figure 8.1. Starting from some arbitrary point A_i, whose
Cartesian coordinates are x_i, y_i, let curve c take us to point A_{i+1}, of
coordinates x_{i+1}, y_{i+1}. In so doing, radius r_i, which connects origin
O to point A_i, and whose inclination with respect to the x-axis is ϑ_i
radians, is multiplied by a factor k, following an anticlockwise rotation
through the *elementary angular increment* φ. We write

$$x_{i+1} = kr_i \cos(\vartheta_i + \varphi) = kr_i(\cos \vartheta_i \cos \varphi - \sin \vartheta_i \sin \varphi),$$
$$y_{i+1} = kr_i \sin(\vartheta_i + \varphi) = kr_i(\sin \vartheta_i \cos \varphi + \cos \vartheta_i \sin \varphi), \quad (8.1)$$

and since

$$x_i = r_i \cos \vartheta_i, \qquad y_i = r_i \sin \vartheta_i,$$

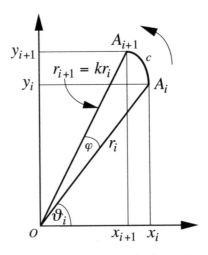

Figure 8.1. Rotation with elongation.

we get

$$x_{i+1} = k(x_i \cos \varphi - y_i \sin \varphi),$$
$$y_{i+1} = k(x_i \sin \varphi + y_i \cos \varphi). \tag{8.2}$$

This can be written in matrix form

$$\begin{bmatrix} x_{i+1} \\ y_{i+1} \end{bmatrix} = k \begin{bmatrix} \cos \varphi & -\sin \varphi \\ \sin \varphi & \cos \varphi \end{bmatrix} \begin{bmatrix} x_i \\ y_i \end{bmatrix} \tag{8.3}$$

and is known as the *rotation matrix*.

Following n increments identical to φ, with constant k, coordinates x_{i+n}, y_{i+n} become

$$\begin{bmatrix} x_{i+n} \\ y_{i+n} \end{bmatrix} = k^n \begin{bmatrix} \cos \varphi & -\sin \varphi \\ \sin \varphi & \cos \varphi \end{bmatrix}^n \begin{bmatrix} x_i \\ y_i \end{bmatrix}.$$

It can be shown by induction (or using the Cayley-Hamilton method, for instance) that the nth power of the matrix is given by

$$\begin{bmatrix} \cos \varphi & -\sin \varphi \\ \sin \varphi & \cos \varphi \end{bmatrix}^n = \begin{bmatrix} \cos n\varphi & -\sin n\varphi \\ \sin n\varphi & \cos n\varphi \end{bmatrix}, \tag{8.4}$$

and we obtain, putting $i = 0$,

$$\begin{bmatrix} x_n \\ y_n \end{bmatrix} = k^n \begin{bmatrix} \cos n\varphi & -\sin n\varphi \\ \sin n\varphi & \cos n\varphi \end{bmatrix} \begin{bmatrix} x_0 \\ y_0 \end{bmatrix}. \tag{8.5}$$

In other words,

$$\begin{aligned} x_n &= k^n(x_0 \cos n\varphi - y_0 \sin n\varphi), \\ y_n &= k^n(x_0 \sin n\varphi + y_0 \cos n\varphi). \end{aligned} \tag{8.6}$$

Equations (8.6) give coordinates x_n, y_n following n increments identical to φ, beginning with initial coordinates x_0, y_0. When $k = 1$, the figure corresponding to equations (8.6) is a circle of radius $\sqrt{x_0^2 + y_0^2}$.

THE MONOGNOMONIC SPIRAL

If point A_0 lies on the x axis, $y_0 = 0$ and equations (8.6) become

$$x_n = k^n x_0 \cos n\varphi, \qquad y_n = k^n x_0 \sin n\varphi$$

hence

$$r_n = \sqrt{x_n^2 + y_n^2} = k^n r_0 \sqrt{\cos^2 n\varphi + \sin^2 n\varphi} = k^n r_0. \tag{8.7}$$

We also have

$$\frac{y_n}{x_n} = \tan \vartheta_n = \tan n\varphi, \quad \text{hence } \vartheta_n = n\varphi. \tag{8.8}$$

Using elementary trigonometry, the above results signify that if a rotation by some incremental angle φ multiplies the radius by k, then n such rotations multiply that radius by k^n, no matter where the process begins.

We can now take an important step, which takes us from the realm of the *discrete* to that of the *continuous,* and state, by virtue of equations (8.7) and (8.8), that rotation by an arbitrary angle ϑ multiplies the radius by $k^{\vartheta/\varphi}$. Whereas the symbol r_n represented the radius following a rotation by angle $n\varphi$, let symbol $r(\vartheta)$ represent the radius following a rotation by angle ϑ with respect to the x-axis. We get, with $r(0) = r_0$,

$$r(\vartheta) = r_0 k^{\vartheta/\varphi},$$

and if we put

$$\mu = \frac{\log_{10}k}{\varphi}, \quad \text{that is,} \quad k = 10^{\mu\varphi}, \tag{8.9}$$

we can write

$$r(\vartheta) = r_0 \times 10^{\mu\vartheta}. \tag{8.10}$$

If we choose any *base* other than 10, the same reasoning applies. In particular, we can write

$$r(\vartheta) = r_0 e^{\lambda\vartheta}, \quad \text{where} \quad \lambda = \frac{\ln k}{\varphi}. \tag{8.11}$$

The polar coordinate equation of the monognomonic spiral defines radius r as a function of angle ϑ and is simply written $r = r_0 e^{\lambda\vartheta}$, hence the name *logarithmic spiral*.

Clearly, the locus of the successive diagonal vertices of a whorled monognomonic rectangle is a monognomonic spiral. We will refer to it as the rectangular monognomonic spiral of order m. Given m, the *flare coefficient* λ_m can be calculated for a *left-handed spiral* (one that uncoils in the anticlockwise direction) by observing that radius r is multiplied by a factor Φ_m following an anticlockwise rotation of $\pi/2$; in other words,

$$r\left(\frac{\pi}{2}\right) = r_0 \Phi_m = r_0 e^{\lambda_m \pi/2}, \qquad \lambda_m = \frac{2 \ln \Phi_m}{\pi}. \tag{8.12}$$

For a right-handed spiral, the radius is multiplied by a factor $1/\Phi_m$ following an anticlockwise rotation of $\pi/2$, or it is multiplied by a factor Φ_m following a clockwise rotation of $\pi/2$; the value of the flare coefficient is the same, albeit with a negative sign.

For the spiral corresponding to the golden rectangle,

$$\lambda_1 = \frac{2}{\pi} \ln \frac{1 + \sqrt{5}}{2} \approx 0.306349,$$

and for the homognomonic rectangle,

$$\lambda_{1/\sqrt{2}} = \frac{2}{\pi} \ln \sqrt{2} \approx 0.2206356.$$

The same reasoning may be applied to whorled figures other than the rectangle. For instance, the spiral corresponding to the golden triangle is such that its radius increases by a factor Φ_m for every 108° rotation. Therefore, $\lambda = 5 \ln(1.618034\ldots)/3\pi = 0.2252908$. For the homognomonic triangle $\lambda = 0.4412712$.

Let us now turn to the matrix

$$\begin{bmatrix} 1 & -\tau \\ \tau & 1 \end{bmatrix}$$

where τ is some arbitrary number. Putting $\tau = \tan \varphi$, we write

$$\begin{bmatrix} 1 & -\tau \\ \tau & 1 \end{bmatrix} = \begin{bmatrix} 1 & -\tan \varphi \\ \tan \varphi & 1 \end{bmatrix} = \frac{1}{\cos \varphi} \begin{bmatrix} \cos \varphi & -\sin \varphi \\ \sin \varphi & \cos \varphi \end{bmatrix},$$

from which it follows that matrix equation

$$\begin{bmatrix} x_{i+1} \\ y_{i+1} \end{bmatrix} = \frac{k}{\sqrt{1+\tau^2}} \begin{bmatrix} 1 & -\tau \\ \tau & 1 \end{bmatrix} \begin{bmatrix} x_i \\ y_i \end{bmatrix} \qquad (8.13a)$$

entails solution

$$\begin{bmatrix} x_n \\ y_n \end{bmatrix} = k^n \begin{bmatrix} \cos n\varphi & -\sin n\varphi \\ \sin n\varphi & \cos n\varphi \end{bmatrix} \begin{bmatrix} x_0 \\ y_0 \end{bmatrix}, \quad \text{with } \tan \varphi = \tau. \quad (8.13b)$$

Comparing with matrix equation (8.5), equations (8.13) both correspond to a spiral whose radius expands by a factor k with every rotation by an angle $\varphi = \tan^{-1}\tau$.

In equation (8.13a), the coefficient $\cos \varphi = 1/\sqrt{1+\tau^2}$ is a function of the coarseness of the chosen angular resolution and tends to 1 with decreasing values of φ. We shall call it the *sampling factor* corresponding to *angular resolution* φ. For example, if one wishes to draw a circle of

155

radius r on a computer screen, with an angular resolution of φ radians, a simple algorithm is

$$\begin{bmatrix} x_{i+1} \\ y_{i+1} \end{bmatrix} = \frac{1}{\sqrt{1 + \tau^2}} \begin{bmatrix} 1 & -\tau \\ \tau & 1 \end{bmatrix} \begin{bmatrix} x_i \\ y_i \end{bmatrix}, \quad \text{with } x_0 = r, y_0 = 0, \tau = \tan \varphi.$$

(8.14)

For the rectangular monognomonic spiral of order m, let $k_{m,\varphi}$ be the value of the coefficient k corresponding to angular increment φ. Putting $y_0 = 0$, $x_0 = r_0$,

$$\Phi_m = \frac{r\left(\dfrac{\pi}{2}\right)}{r_0} = (k_{m,\varphi})^{\pi/2\varphi}, \qquad k_{m,\varphi} = e^{\lambda_m \varphi}.$$

(8.15)

The finite difference equations of the rectangular monognomonic spiral of order m, with angular increment $\varphi = \tan^{-1} \tau$, are indifferently

$$\begin{bmatrix} x_{i+1} \\ y_{i+1} \end{bmatrix} = \frac{e^{\lambda_m \tan^{-1} \tau}}{\sqrt{1 + \tau^2}} \begin{bmatrix} 1 & -\tau \\ \tau & 1 \end{bmatrix} \begin{bmatrix} x_i \\ y_i \end{bmatrix},$$

(8.16a)

$$\begin{bmatrix} x_{i+1} \\ y_{i+1} \end{bmatrix} = e^{\lambda_m \varphi} \begin{bmatrix} \cos \varphi & -\sin \varphi \\ \sin \varphi & \cos \varphi \end{bmatrix} \begin{bmatrix} x_i \\ y_i \end{bmatrix},$$

(8.16b)

$$\begin{bmatrix} x_n \\ y_n \end{bmatrix} = e^{\lambda_m n \varphi} \begin{bmatrix} \cos n\varphi & -\sin n\varphi \\ \sin n\varphi & \cos n\varphi \end{bmatrix} \begin{bmatrix} x_0 \\ y_0 \end{bmatrix}.$$

(8.16c)

Its Cartesian-coordinate equations, in terms of parameter ϑ, are

$$\begin{bmatrix} x \\ y \end{bmatrix} = e^{\lambda_m \varphi} \begin{bmatrix} \cos \vartheta & -\sin \vartheta \\ \sin \vartheta & \cos \vartheta \end{bmatrix} \begin{bmatrix} x_0 \\ y_0 \end{bmatrix}.$$

(8.16d)

Its polar-coordinate equation, expressing radius r as a function of angle ϑ, is

$$r = r_0 e^{\lambda_m \vartheta}$$

(8.16e)

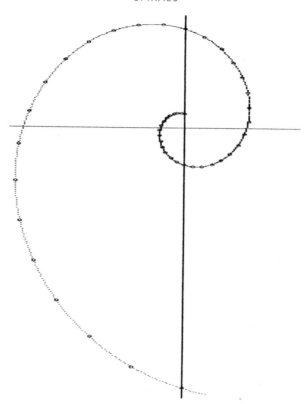

Figure 8.2. The golden rectangular monognomonic spiral, showing sampling points.

with

$$\lambda_m = \frac{2 \ln \Phi_m}{\pi} \quad \Phi_m = \frac{m + \sqrt{4 + m^2}}{2}.$$

Observe that $e^{\lambda_m \varphi}$ is the elongation of radius r following a rotation equal to elementary increment φ, and $\cos \varphi = \sqrt{1 + \tau^2}$ is the sampling factor.

It is important to observe that all of equations (8.16) correspond to exactly the same curve. Whereas (8.16d) and (8.16e) apply to *every* point on the curve, equations (8.16a, b, c) represent an infinite subset of points separated by discrete (noninfinitesimal) intervals. Iterative equations (8.16a, b) define every such point in terms of the preceding point, whereas equation (8.16c) furnishes an explicit formulation in terms of the total number of iterations. Figure 8.2 represents a golden rectangular monognomonic spiral upon which can be found, represented by little beads, the sampling points corresponding to $\varphi = 2\pi/30$ radians, that is, $\tau = 0.2125566$, $e^{\lambda_m \varphi} = 1.066265$ in

$$\begin{bmatrix} x_{i+1} \\ y_{i+1} \end{bmatrix} = 1.066265 \begin{bmatrix} 1 & -0.2125566 \\ 0.2125566 & 1 \end{bmatrix} \begin{bmatrix} x_i \\ y_i \end{bmatrix}. \qquad (8.17a)$$

If the chosen value of φ is very small, we get

$$k_{m,\varphi} = e^{\lambda_m \varphi} = (1 + \lambda_m \varphi) \quad \text{and} \quad \cos \varphi \approx 1,$$

and for a right-handed spiral

$$k_{m,\varphi} = e^{-\lambda_m \varphi} = (1 - \lambda_m \varphi).$$

With $\tau = 0.001$, we obtain the following iterative equations for the left-handed golden rectangular spiral:

$$\begin{bmatrix} x_{i+1} \\ y_{i+1} \end{bmatrix} = 1.000306 \begin{bmatrix} 1 & -0.001 \\ 0.001 & 1 \end{bmatrix} \begin{bmatrix} x_i \\ y_i \end{bmatrix}. \qquad (8.17b)$$

Again, equations (8.17) correspond to exactly the same spiral (provided that x_0 and y_0 are the same for both); only the sampling rate differs.

Self-similarity

Imagine that a photocopy is made of the logarithmic spiral with an enlargement factor k. The equation of the new spiral is $r' = kr_0 e^{\lambda \vartheta}$, which can be rewritten $r' = r_0 e^{\lambda \vartheta} e^{\lambda \theta} = r_0 e^{\lambda(\varphi + \theta)}$, where $\theta = (\ln k)/\lambda$.

The enlarged spiral is therefore identical to the original, albeit rotated anticlockwise by angle θ. The reader may make such an enlargement on transparent paper, then apply it over the original, pole over pole. Upon rotating the transparent sheet, he or she will come upon some angle θ, for which total coincidence is obtained. If the photocopier's enlargement factor is known, the reader will be able to determine flare coefficient λ, and vice versa.

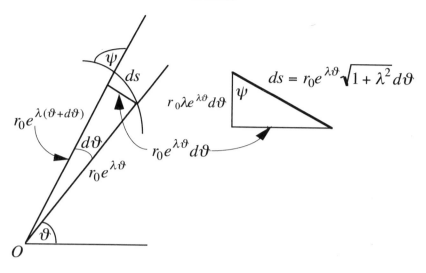

Figure 8.3. Equiangularity.

Equiangularity

Figure 8.3 shows a logarithmic spiral segment inscribed between two values of the radius, following rotations ϑ and $\vartheta + d\vartheta$ with respect to the horizontal axis, upon which the radius is r_0. Making $e^x \approx (1 + x)$ for very small values of x, the difference between the two radii is

$$r_0 e^{\lambda(\vartheta + d\vartheta)} - r_0 e^{\lambda\vartheta} = r_0 e^{\lambda\vartheta}(e^{\lambda d\vartheta} - 1) = r_0 \lambda e^{\lambda\vartheta}\, d\vartheta. \quad (8.18)$$

If ψ denotes the angle between the tangent to the spiral curve at a given point and the radius at that point, we have

$$\cot\psi = \frac{r_0 \lambda e^{\lambda\vartheta}\, d\vartheta}{r_0 e^{\lambda\vartheta}\, d\vartheta} = \lambda. \quad (8.19)$$

Angle ψ is therefore constant. Because of that peculiar property of the logarithmic spiral, it is often referred to as the *equiangular* spiral, a name introduced by Roger Cotes in 1722 in his *Harmonia Mensurarum,* following the curve's description by Descartes in his 1638 letters to Mersenne.[1] Obviously, the limiting forms of the spiral are the circle, for which $\cot^{-1}\lambda = \pi/2$, that is, $\lambda = 0$, and the straight line, for which $\cot^{-1}\lambda = 0$, that is, $\lambda = \infty$.

If you hold a toy periscope horizontally to your eye after tilting the mirror that is farthest from you by a small angle with respect to its normal 90° inclination, and endeavor to walk always toward a fixed

[1] These can be found in Adam and Tannery's *Oeuvres de Descartes* (Paris, 1898).

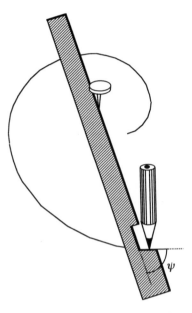

Figure 8.4. Constructing an equiangular spiral. After Martin Gardner, *The Unexpected Hanging* (New York: Simon and Schuster, 1969), p. 108.

point in the room, as seen through the periscope, your friends will observe that you are moving in a spiral that progressively closes in on that object, as it wraps itself around the object. According to D'Arcy Thompson, certain insects, owing to the compound structure of their eye, see objects abeam, which causes them to move in spirals as they head toward them.

Based on that property of equiangularity, Martin Gardner devised an ingenious instrument for drawing equiangular spirals, shown in figure 8.4. In his words, "angle [ψ] may be of any size between 0 and 180 degrees. By keeping one edge of the strip on the spiral's pole and ruling short line segments along the oblique straightedge as this straightedge is moved toward or away from the pole, you produce a series of chords of the spiral in much the same way the Epeira spins its web. The device ensures that all these chords cut the radius vector at the same angle."[2]

Imagine an airplane that takes off at some point on the globe and endeavors to always cut the earth's meridians at some constant angle. If that angle is 0°, the airplane will circle the globe forever, going from

[2] *The Unexpected Hanging*, p. 108.

Figure 8.5. Escher's Loxodrome. (M. C. Escher, *Sphere Spirals*. © 1998 Cordon Art B.V.–Baarn–Holland. All rights reserved.)

pole to pole, along whatever longitude it took off from. If the angle is 90°, the plane will travel along a fixed latitude. If the angle is neither 0 nor 90°, the airplane's trajectory will eventually "wind up" at a pole, following an infinite number of ever smaller rotations around it. The length of that trajectory, called the *loxodrome,* is finite! One cannot help but marvel at Maurits Escher's extraordinary premonition of infinite processes and self-similarity, as illustrated in figure 8.5, showing an array of loxodromes, which Escher intuitively called *spherical spirals.* All bands and intervals between bands are identical.

Perimeter of the Spiral

The hypotenuse ds of the little triangle in figure 8.3 is a segment of the spiral itself:

$$ds = \sqrt{(r_0 e^{\lambda \vartheta} d\vartheta)^2 + (r_0 \lambda e^{\lambda \vartheta} d\vartheta)^2} = r_0 \sqrt{1 + \lambda^2} e^{\lambda \vartheta} d\vartheta.$$

In order to calculate the spiral's perimeter between some initial point

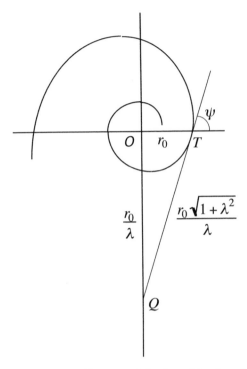

Figure 8.6. Rectification of the logarithmic spiral.

and the pole, it is necessary to integrate segment ds from $\vartheta = 0$ to $-\infty$. That perimeter is

$$S = \int_0^{-\infty} r_0\sqrt{1 + \lambda^2}\,e^{\lambda\vartheta}\,d\vartheta = \frac{r_0\sqrt{1 + \lambda^2}}{\lambda}. \qquad (8.20)$$

Turning to figure 8.6, this equation means that the perimeter of the spiral from point T, where the radius is r_0, is equal to the length of tangent TQ at that point. That property is referred to as the *rectification of the logarithmic spiral*. Remembering that the angle ψ between the radius and the tangent is equal to $\cot^{-1}\lambda$, we have

$$S = \frac{r_0}{\cos\psi}. \qquad (8.21)$$

Surprising as that may seem, the logarithmic spiral is *finite* in length, despite the infinite nature of the winding process that takes it from some arbitrary radius to the pole, where its radius is zero. That fact was discovered in 1645 by Evangelista Torricelli, famous for his invention of the barometer.[3] This paradox is akin to that of Achilles and the tortoise, in that the sum of an infinite number of geometrically shrinking distances converges to a finite quantity. Every high school student has been told, at some time or another, the tale of Achilles and the tortoise, one of the four paradoxes of Zeno the Eleate, a student of Parmenides who lived in the fifth century B.C. Achilles runs twice as fast as the tortoise and challenges her to a race, granting her a head start of, say, 10 meters. By the time the swift Achilles covers that initial distance, the slow but unrelenting tortoise has covered 5 meters. By the time Achilles covers that distance, the tortoise has covered 2.5 meters, and so on. After all eternity has elapsed, the tortoise remains half of some residual distance ahead, and Achilles never catches up.

Indeed, starting with some arbitrary chord of length c, the next chord measures c/Φ_m. The following chords measure c/Φ_m^2, c/Φ_m^3, ..., and the sum of the infinite series is $c/(\Phi_m - 1)$, a finite number. As an exercise, the reader may calculate the flare coefficient of the spiral that is totally inscribed within a given circle, which it intersects at one point, and whose perimeter is equal to the circle's circumference. The answer is

$$\lambda = \frac{1}{\sqrt{4\pi^2 - 1}} \approx 0.1612098.$$

This corresponds to $\psi \approx 80°50'31''$. Figure 8.7 shows that spiral, whose harmony is indeed singular. Bernoulli discovered another singular spiral, which is identical to its own involute. That spiral may be shown to possess a flare coefficient $\lambda \approx 0.2745$, that is, $\psi \approx 74°39'$. As a further exercise, the reader may construct the interlocking pattern of figure 8.8, consisting of logarithmic spirals whose radius doubles with every 60° rotation, corresponding to $(3 \ln 2)/\pi$, and whose poles are located upon the vertices of a lattice of equilateral triangles.

---

[3] Torricelli was a student of Galileo who advocated in his *Discorsi et dimostrazioni mathematiche intorno, a due nove scienze* (1638) that despite Aristotle's assertions that Nature abhors a vacuum, man could actually create a vacuum in the laboratory.

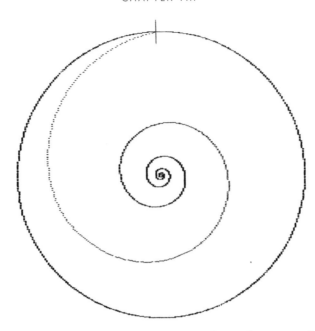

Figure 8.7. This spiral's length is equal to the circle's circumference.

Figure 8.8. Interlocking spirals.

THE RECTANGULAR DIGNOMONIC SPIRAL

We can now advance one step further and examine the matrix

$$
\begin{bmatrix}
\cos \varphi & -\dfrac{1}{a}\sin \varphi \\
a \sin \varphi & \cos \varphi
\end{bmatrix}.
$$

As an exercise, the reader may prove by induction that

$$
\begin{bmatrix}
\cos \varphi & -\dfrac{1}{a}\sin \varphi \\
a \sin \varphi & \cos \varphi
\end{bmatrix}^{n}
=
\begin{bmatrix}
\cos n\varphi & -\dfrac{1}{a}\sin n\varphi \\
a \sin n\varphi & \cos n\varphi
\end{bmatrix}.
\tag{8.22}
$$

Consider now the matrix

$$
\begin{bmatrix}
1 & -\dfrac{r}{m}\tau \\
\dfrac{s}{m}\tau & 1
\end{bmatrix},
\tag{8.23}
$$

where r, m are positive numbers, and $m = \sqrt{rs}$. Putting $\tau = \tan \varphi$, it follows that matrix equation

$$
\begin{bmatrix} x_{i+1} \\ y_{i+1} \end{bmatrix}
= \frac{k}{\sqrt{1 + \tau^2}}
\begin{bmatrix}
1 & -\dfrac{r}{m}\tau \\
\dfrac{s}{m}\tau & 1
\end{bmatrix}
\begin{bmatrix} x_i \\ y_i \end{bmatrix}
\tag{8.24a}
$$

entails solution

$$
\begin{bmatrix} x_n \\ y_n \end{bmatrix}
= k^n
\begin{bmatrix}
\cos n\varphi & -\dfrac{r}{m}\sin n\varphi \\
\dfrac{s}{m}\sin n\varphi & \cos n\varphi
\end{bmatrix}
\begin{bmatrix} x_0 \\ y_0 \end{bmatrix},
\quad \text{with } \varphi = \tan^{-1}\tau.
$$

$$
\tag{8.24b}
$$

If we make $y_0 = 0$, we get

$$x_n = x_0 k^n \cos n\varphi, \qquad y_n = x_0 \frac{s}{m} k^n \sin n\varphi, \qquad (8.25)$$

$$\frac{y_n}{x_n} = \tan \vartheta_n = \frac{s}{m} \tan n\varphi. \qquad (8.26)$$

Whereas in the case of the monognomonic spiral, $\vartheta_n = n\varphi$ for all values of n, this is true of equation (8.26) only when ϑ_n is a multiple of $\pi/2$. If we put

$$\frac{y\left(\dfrac{\pi}{2}\right)}{x_0} = \frac{s}{m} k^{\frac{\pi}{2\varphi}} = \phi_s, \qquad (8.27)$$

this yields

$$k^{\frac{\pi}{\varphi}} = \frac{m^2}{s^2} \phi_s^2 = \frac{r}{s} \phi_s^2 = \phi_r \phi_s. \qquad (8.28)$$

On the other hand,

$$\frac{x(\pi)}{x_0} = k^{\frac{\pi}{\varphi}} \cos(\pi) = -k^{\frac{\pi}{\varphi}} \quad \text{hence} \quad \frac{x(\pi)}{x_0} = -\phi_r \phi_s. \qquad (8.29)$$

The curve corresponding to matrix equations (8.24) is that of a dignomonic spiral of proportions ϕ_s, ϕ_r, as in figure 8.9. Remembering that

$$m = \sqrt{rs}, \qquad \phi_r \phi_s = \Phi_m^2, \qquad \frac{\phi_r}{\phi_s} = \frac{r}{s},$$

equation (8.28) yields

$$k^{\frac{\pi}{2\varphi}} = \sqrt{\phi_r \phi_s} = \Phi_m.$$

Comparing with equation (8.15), $k = k_{m,\varphi}$, and the finite-difference equations of the rectangular dignomonic spiral are

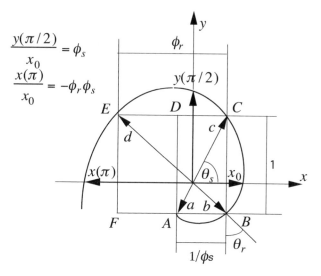

$$\frac{y(\pi/2)}{x_0} = \phi_s$$

$$\frac{x(\pi)}{x_0} = -\phi_r\phi_s$$

Figure 8.9. The dignomonic spiral.

$$\begin{bmatrix} x_{i+1} \\ y_{i+1} \end{bmatrix} = \frac{e^{\lambda_m \tan^{-1}\tau}}{\sqrt{1+\tau^2}} \begin{bmatrix} 1 & -\dfrac{r}{m}\tau \\ \dfrac{s}{m}\tau & 1 \end{bmatrix} \begin{bmatrix} x_i \\ y_i \end{bmatrix}, \qquad (8.30a)$$

$$\begin{bmatrix} x_{i+1} \\ y_{i+1} \end{bmatrix} = e^{\lambda_m \varphi} \begin{bmatrix} \cos\varphi & -\dfrac{r}{m}\sin\varphi \\ \dfrac{s}{m}\sin\varphi & \cos\varphi \end{bmatrix} \begin{bmatrix} x_i \\ y_i \end{bmatrix}, \qquad (8.30b)$$

$$\begin{bmatrix} x_n \\ y_n \end{bmatrix} = e^{\lambda_m n\varphi} \begin{bmatrix} \cos n\varphi & -\dfrac{r}{m}\sin n\varphi \\ \dfrac{s}{m}\sin n\varphi & \cos n\varphi \end{bmatrix} \begin{bmatrix} x_0 \\ y_0 \end{bmatrix}. \qquad (8.30c)$$

Its analytical equations, in terms of variable parameter ρ, are

$$\begin{bmatrix} x \\ y \end{bmatrix} = e^{\lambda_m \rho} \begin{bmatrix} \cos\rho & -\dfrac{r}{m}\sin\rho \\ \dfrac{s}{m}\sin\rho & \cos\rho \end{bmatrix} \begin{bmatrix} x_0 \\ y_0 \end{bmatrix}, \text{with } \tau = \tan\varphi, \quad \lambda_m = \frac{2\ln\Phi_m}{\pi}.$$

$$(8.30d)$$

The symbol ρ was chosen instead of the symbol ϑ which appears in equations (8.16c), as only the latter represents angle $\tan^{-1} y/x$.

As an example, a family of dignomonic spirals for which $rs = 1$ can be obtained by assigning various values to α in the equation

$$
\begin{bmatrix} x_{i+1} \\ y_{i+1} \end{bmatrix} = 1.000306 \begin{bmatrix} 1 & -0.001\alpha \\ \dfrac{0.001}{\alpha} & 1 \end{bmatrix} \begin{bmatrix} x_i \\ y_i \end{bmatrix}. \qquad (8.31)
$$

Given rectangles $ABCD$ and $ADEF$ shown in figure 8.9, the corresponding spiral envelope can be drawn as follows. Radii $OA = a$ and $OC = c$ are collinear. Since every rotation by angle π multiplies the radius by factor $\phi_r \phi_s$, we have $AC = a(1 + \phi_r \phi_s)$, and since

$$
AC = \sqrt{1 + \frac{1}{\phi_s^2}},
$$

we get

$$
a = \frac{\sqrt{1 + \phi_s^2}}{\phi_s(1 + \phi_r \phi_s)}. \qquad (8.32)
$$

Relative to origin O, the Cartesian coordinates of points (x_i, y_i) on the spiral are given by matrix equations (8.30), where

$$
x_0 = -a \cos \theta_s = -\frac{1}{\phi_s(1 + \phi_r \phi_s)}, \qquad y_0 = -a \sin \theta_s = -\frac{1}{1 + \phi_r \phi_s}.
$$

$$
(8.33)
$$

We can now plot, for every value of i, spiral coordinates x', y' with respect to origin A

$$
x_i' = x_i + a \cos \theta_s, \quad y_i' = y_i + a \sin \theta_s,
$$

where $(a \cos \theta_s, a \sin \theta_s)$ are the Cartesian coordinates of pole O with respect to origin A.

THE ARCHIMEDEAN SPIRAL

A disturbing paradox is offered by another well-known, though less miraculous, spiral, namely, the *equable* spiral, otherwise known

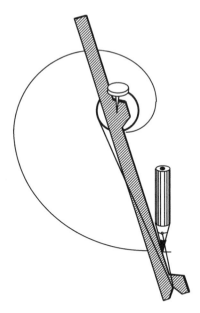

Figure 8.10. Constructing an Archimedean spiral. After Gardner, *The Unexpected Hanging,* p. 105.

as *Archimedean.* Whereas the radius of the logarithmic spiral grows *geometrically* with the rotation angle, that of the Archimedean spiral grows or shrinks *linearly.* Its equation is $r = k\theta$, and it can be shown that its perimeter tends to infinity as it wraps itself around the pole. That property is akin to that of the *harmonic* series $1 + 1/2 + 1/3 + 1/4, \ldots$, whose terms are evanescent, but which diverges. That sum was proven to be infinite by Bernoulli's brother Jonn, and published by Jakob in 1713, giving credit to his brother. The Archimedean spiral is materialized by the trajectory of an object that moves radially at constant speed toward or away from the center of a surface that rotates with uniform angular speed, in much the same way a gramophone needle moves on the surface of a 33 RPM record. Archimedean spirals can be approached by unwinding a coil of string wrapped around a drum: the resulting spiral, known an the *involute* spiral, is not really Archimedean, for it is not the radius, but the length of the tangent to the drum that increases linearly with the rotation angle. Ingenious instruments for constructing the involute and Archimedean spirals can also be found in Martin Gardner's *The Unexpected Hanging* (fig. 8.10).

The Archimedean spiral was celebrated by the ancient Greeks be-

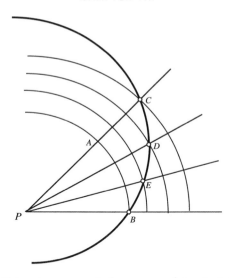

Figure 8.11a. Trisecting an angle by means of the Archimedean spiral.

cause it offered pseudo-solutions to the problem of squaring the circle as well as that of trisecting an angle, as in figure 8.11a. Assuming that there exists an exact method for constructing an Archimedean spiral, it is required to trisect angle *CPB* in figure 8.11b. An Archimedean spiral segment *CB* is constructed, whose pole is *P*. Following that initial construction, circle arc *AB* is drawn, and line segment *AC* is trisected in the classical manner. Circular arcs are then drawn from the dividing points, which intersect the spiral at *D* and *E*. Finally, lines *PD* and *PE* are drawn, which can be shown to trisect angle *CPB*.

Figure 8.11b shows how a circle can be squared. An Archimedean spiral that starts at pole *O* completes its first revolution at point *T*. Tangent *AT* is drawn at that point, which intersects at *A* the perpendicu-

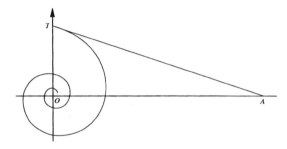

Figure 8.11b. Squaring of the circle by means of the Archimedean spiral.

lar erected upon OT from pole O. It can be shown that $OA = \pi\, OT$. Since the problem of squaring the circle essentially boils down to determining π, the method is seen to provide some answer. The above two constructions violate, however, the basic stipulation that the problems be solved using only straightedge and compass, an impossible condition to satisfy. The methods, no matter how elegant, have to be dismissed as bona fide solutions to Plato's problems, along with thousands of others proposed over the centuries.

DAMPED OSCILLATIONS

Next we examine a generalization of the above results that will prove particularly relevant upon studying simple harmonic motion, and the behavior of oscillating electrical circuits. In the appendix, we shall take the opportunity to introduce the reader to an important result of the *calculus of finite differences,* which we could, of course, have used from the outset of this chapter. We preferred instead to resort to a step-by-step approach, in the hope of making the subject more easily accessible. We first introduce the following fundamental matrix:

$$\begin{bmatrix} \cos\varphi + b\sin\varphi & -c\sin\varphi \\ a\sin\varphi & \cos\varphi - b\sin\varphi \end{bmatrix}, \quad \text{where } b = \sqrt{ac-1}.$$

$$(8.34)$$

It can be easily shown by induction that

$$\begin{bmatrix} \cos\varphi + b\sin\varphi & -c\sin\varphi \\ a\sin\varphi & \cos\varphi - b\sin\varphi \end{bmatrix}^{n} =$$

$$\begin{bmatrix} \cos n\varphi + b\sin n\varphi & -c\sin n\varphi \\ a\sin n\varphi & \cos n\varphi - b\sin n\varphi \end{bmatrix}, \quad (8.35)$$

from which it follows that the equation

$$\begin{bmatrix} x_{i+1} \\ y_{i+1} \end{bmatrix} = \begin{bmatrix} \cos\varphi + b\sin\varphi & -c\sin\varphi \\ a\sin\varphi & \cos\varphi - b\sin\varphi \end{bmatrix} \begin{bmatrix} x_i \\ y_i \end{bmatrix} \quad (8.36)$$

entails solution

$$\begin{bmatrix} x_n \\ y_n \end{bmatrix} = \begin{bmatrix} \cos n\varphi + b \sin n\varphi & -c \sin n\varphi \\ a \sin n\varphi & \cos n\varphi - b \sin n\varphi \end{bmatrix} \begin{bmatrix} x_0 \\ y_0 \end{bmatrix}. \tag{8.37}$$

Matrix 8.34 can be written

$$\cos \varphi \begin{bmatrix} 1 + b \tan \varphi & -c \tan \varphi \\ a \tan \varphi & 1 - b \tan \varphi \end{bmatrix}. \tag{8.38}$$

Putting

$$a = \frac{\lambda}{\omega}, \qquad b = \frac{\rho}{\omega}, \qquad c = \frac{\gamma}{\omega}, \tag{8.39}$$

where γ, λ, ρ, and ω are real positive numbers, we get, using matrix (8.34),

$$\omega = \sqrt{\gamma\lambda - \rho^2}. \tag{8.40}$$

If we now introduce the "elementary time increment" τ, such that

$$\tan \varphi = \frac{\omega\tau}{1 - \rho\tau} \tag{8.41}$$

we get

$$1 + b \tan \varphi = \frac{1}{1 - \rho\tau}, \qquad 1 - b \tan \varphi = \frac{1 - 2\rho\tau}{1 - \rho\tau}. \tag{8.42}$$

Matrix (8.38) now becomes

$$\frac{\cos \varphi}{1 - \rho\tau} \begin{bmatrix} 1 & -\gamma\tau \\ \lambda\tau & 1 - 2\rho\tau \end{bmatrix}, \tag{8.43}$$

from which it follows that matrix equation

$$\begin{bmatrix} x_{i+1} \\ y_{i+1} \end{bmatrix} = \begin{bmatrix} 1 & -\gamma\tau \\ \lambda\tau & 1 - 2\rho\tau \end{bmatrix} \begin{bmatrix} x_i \\ y_i \end{bmatrix} \tag{8.44}$$

entails solution

$$
\begin{bmatrix} x_n \\ y_n \end{bmatrix} = \left(\frac{1 - \rho\tau}{\cos \varphi} \right)^n \begin{bmatrix} \cos n\varphi + \dfrac{\rho}{\omega} \sin n\varphi & -\dfrac{\gamma}{\omega} \sin n\varphi \\[3mm] \dfrac{\lambda}{\omega} \sin n\varphi & \cos n\varphi - \dfrac{\rho}{\omega} \sin n\varphi \end{bmatrix} \begin{bmatrix} x_0 \\ y_0 \end{bmatrix},
$$

(8.45)

with

$$
\omega = \sqrt{\gamma\lambda - \rho^2}, \qquad \varphi = \tan^{-1} \frac{\omega\tau}{1 - \rho\tau}.
$$

The above statement should be completed with the phrase "and conversely." Both matrix equations actually constitute equivalent statements, the first of which is iterative, and the second an explicit function of n.

Remembering that for small values of τ we have $(1 - \rho\tau)^n = e^{-\rho n\tau}$, $\cos \varphi = 1$ and $\varphi = \tan \varphi = \omega\tau$, expression (8.45) becomes

$$
\begin{bmatrix} x_n \\ y_n \end{bmatrix} = e^{-\rho n\tau} \begin{bmatrix} \cos \omega n\tau + \dfrac{\rho}{\omega} \sin \omega n\tau & -\dfrac{\gamma}{\omega} \sin \omega n\tau \\[3mm] \dfrac{\lambda}{\omega} \sin \omega n\tau & \cos \omega n\tau - \dfrac{\rho}{\omega} \sin \omega n\tau \end{bmatrix} \begin{bmatrix} x_0 \\ y_0 \end{bmatrix}.
$$

Putting $t = n\tau$, and letting symbols x, y without indices represent *continuous* variables x, y as functions of *continuous* parameter t, we obtain the analytical expressions

$$
x = e^{-\rho t} \left(x_0 \left(\cos \omega t + \frac{\rho}{\omega} \sin \omega t \right) - y_0 \frac{\gamma}{\omega} \sin \omega t \right),
$$

(8.46a)

$$
y = e^{-\rho t} \left(x_0 \frac{\lambda}{\omega} \sin \omega t + y_0 \left(\cos \omega t - \frac{\rho}{\omega} \sin \omega t \right) \right), \quad \text{with } \omega = \sqrt{\gamma\lambda - \rho^2}.
$$

(8.46b)

Equations (8.46) can equally be written

$$
x = e^{-\rho t} \left(x_0 \cos \omega t + \frac{\rho x_0 - \gamma y_0}{\omega} \sin \omega t \right),
$$

(8.47a)

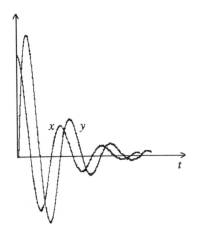

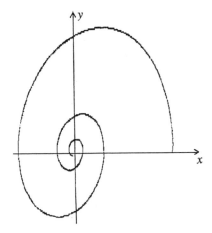

Figure 8.12a. Curves of x and y against time t.

Figure 8.12b. Phase portrait of x, y.

$$y = e^{-\rho t} \left(y_0 \cos \omega t + \frac{\lambda x_0 - \rho y_0}{\omega} \sin \omega t \right). \qquad (8.47b)$$

When ρ is very small,

$$x = e^{-\rho t} \left(x_0 \cos \omega t - y_0 \sqrt{\frac{\gamma}{\lambda}} \sin \omega t \right),$$

$$y = e^{-\rho t} \left(x_0 \sqrt{\frac{\lambda}{\gamma}} \sin \omega t + y_0 \cos \omega t \right), \text{ with } \omega = \sqrt{\gamma \lambda}. \qquad (8.48)$$

Figure 8.12a shows the curves of x and y against time t, highlighting the damped oscillatory nature of both variables. Figure 8.12b shows y as a function of x. The resulting spiral is referred to as the *phase portrait* of the observed phenomenon, and the $x - y$ coordinate system is referred to in that case as the *phase space*.

The Simple Pendulum

Perhaps one of the simplest mechanical machines known to man is the simple pendulum (fig. 8.13). This consists of a small, dense body of mass m, hanging at the end of a string of length l. Ideally, the entire mass m is concentrated in a single dimensionless point. The string is

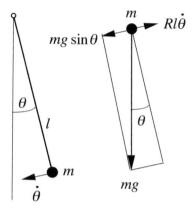

Figure 8.13. The simple pendulum.

weightless and opposes no resistance to the oscillatory motion about rotation axis O. Let θ represent the angular displacement of the string with respect to a vertical reference line, and $\dot{\theta}$ its angular speed, measured in radians per second. We shall assume that the displacement of mass m through air is opposed by a viscous resistance $Rl\dot{\theta}$, which is proportional to the linear tangential velocity $l\dot{\theta}$, where R is the force per unit linear velocity. The resulting tangential force is $mg \sin \theta - Rl\dot{\theta}$, where g is the *acceleration of gravity*. In this discussion, we shall always assume that the angular displacement θ is very small, which allows us to put $\sin \theta = \theta$. The *angular* acceleration to which the mass is subjected is therefore

$$\ddot{\theta} = \frac{mg\theta - Rl\dot{\theta}}{ml} = \frac{g}{l}\theta - \frac{R}{m}\dot{\theta},$$

and if symbol τ represents an *infinitesimal* increment of time, we can write

$$\dot{\theta}_{i+1} = \dot{\theta}_i + \ddot{\theta}_i\, \tau = \dot{\theta}_i + \frac{g}{l}\tau\theta_i - \frac{R}{m}\tau\dot{\theta}_i$$

$$= \frac{g}{l}\tau\theta_i + \left(1 - \frac{R}{m}\right)\dot{\theta}_i.$$

(8.49)

175

We similarly have

$$\theta_{i+1} = \theta_i - \dot{\theta}_i \tau,\tag{8.50}$$

hence the matrix equation

$$\begin{bmatrix} \theta_{i+1} \\ \dot{\theta}_{i+1} \end{bmatrix} = \begin{bmatrix} 1 & -\tau \\ \dfrac{g}{l}\tau & 1 - \dfrac{R}{m}\tau \end{bmatrix} \begin{bmatrix} \theta_i \\ \dot{\theta}_i \end{bmatrix}.\tag{8.51}$$

The pendulum is said to be "underdamped" (see the appendix) if $(g/l) > (R/2m)$, and we can write

$$\omega = \sqrt{\dfrac{g}{l} - \dfrac{R^2}{4m^2}}.\tag{8.52}$$

The solution of equation (8.36) is therefore

$$\begin{bmatrix} \theta \\ \dot{\theta} \end{bmatrix} = e^{-\frac{R}{2m}t} \begin{bmatrix} \cos \omega t + \dfrac{R}{2m\omega}\sin \omega t & -\dfrac{1}{\omega}\sin \omega t \\ \dfrac{g}{l\omega}\sin \omega t & \cos \omega t - \dfrac{R}{2m\omega}\sin \omega t \end{bmatrix} \begin{bmatrix} \theta_0 \\ \dot{\theta}_0 \end{bmatrix}.$$

Putting $\dot{\theta}_0$, meaning that the mass m is dropped with zero initial velocity from its initial angular position, θ_0 radians relative to the vertical, we get

$$\theta = \theta_0 e^{-\frac{R}{2m}t}\left(\cos \omega t + \dfrac{R}{2m\omega}\sin \omega t\right),$$

$$\dot{\theta} = \theta_0 \dfrac{g}{l\omega} e^{-\frac{R}{2m}t}\sin \omega t.\tag{8.53}$$

If friction with air is small, we get

$$\omega = \sqrt{\dfrac{g}{l}}, \qquad \theta = \theta_0 e^{-\frac{R}{2m}t}\cos \omega t, \qquad \dot{\theta} = \theta_0\sqrt{\dfrac{g}{l}}\, e^{-\frac{R}{2m}t}\sin \omega t.\tag{8.54}$$

The pendulum oscillates about the vertical axis with angular velocity

ω in the complex plane (which under no circumstances should be confused with physical angular velocity $\dot{\theta}$). The oscillation period T is

$$T = \frac{2\pi}{\omega} = 2\pi \sqrt{\frac{l}{g}}. \tag{8.55}$$

The oscillation's amplitude decays exponentially with time at the rate of $\exp(-Rt/2m)$.

Equation (8.55) signifies that provided the initial angular displacement is small (l is large), and damping is not excessive, the pendulum's period depends solely on its length (for a fixed geographic location, given the slight variations of g from one location to another). For centuries, pendulum clocks have been predicated on that principle.

The RLC Circuit

In the chapter on ladders, we examined the behavior of an electrical ladder circuit upon being driven by an outside energy source, which translates into applying input voltage v_0 to the circuit. When reactive components are used, namely, the inductance and the capacitance, that *forced* voltage was assumed to be sinusoidal. Here, using a totally different approach, we shall examine the *transient* behavior of an electrical circuit containing reactive components, when it is left to oscillate *freely*.

Again, the components are passive, meaning that they contain no intrinsic electrical energy source. They are linear, meaning that the mathematical expressions characterizing their individual behavior, in terms of the two fundamental physical magnitudes, namely, *current* and potential *difference* are linear. Linearity is basically a simplifying assumption, which allows the superposition, without interaction, of currents generated by different sources within any given component. Linearity also results in frequency conservation. Whenever a linear component is traversed by a current of a given frequency, no other frequencies are generated by the component itself.

In real life, physical components are roughly linear only within a limited range, which leads engineers to consistently endeavor to contain the component's working environment within that range. Any nonlinearity encountered in a component is regarded as a marginal deviation from nominal behavior. In recent years, however, we have witnessed a spectacular reversal of that paradigm, with the systematic study of *chaotic* behavior, which is basically predicated upon nonlinearity. A new paradigm was born, and its followers have embarked upon a passionate search for chaotic behavior not only in electrical circuits but in other phenomena as well, in every discipline known to man, ranging from weather prediction (of which, incidentally, the whole movement was born), to the stock market, heart fibrillation, oscillating chemical reactions, brain waves, and others, to such an extent that the so-called *butterfly effect* has become the trademark of the new trend and the emblem of its sometimes fanatic devotees. Be that as it may, let us prosaically return to our three linear components and attempt a *finite differences* approach to the study, first of their individual, and then of their collective, behavior.

Before we proceed, it is necessary to give a simple definition of the notions of electrical current and potential difference. Electrical current within a component can be visualized as the flow in a given direction within that component, of extremely large quantities of elementary particles, the electrons. Each electron carries the elemental magnitude known as electrical charge. The total charge (measured in *coulombs*) traversing the component during one second, is the electrical current (measured in *amperes*). Electrical potential can be regarded as an electrical force that is potentially capable of *driving* a current through a component. It is often referred to as the *electromotive force.* Potential difference is measured in *volts*. The relationship between the voltage *across,* and the current *within* a component, depends upon the nature of the component and, as we shall see, involves the time element.

The Resistor

The simplest of all electrical components is the resistor, diagramatically shown in figure 8.14. Current is represented by a solid arrow, and potential difference by a dotted arrow standing upon a datum line, which symbolizes the reference point with respect to which the potential difference is measured. Resistance R is measured in ohms. Current I_R,

Figure 8.14. The resistor.

measured in amperes, when driven by a potential difference of V_R volts across the resistor, is given by the very simple (linear) relationship

$$I_R = \frac{V_R}{R}.$$ (8.56)

The Capacitor

Also called a condenser, the *capacitor* consists of a nonconducting *dielectric* thin film, sandwiched between two conducting surfaces, the *electrodes*. It is schematically represented by the diagram in figure 8.15. Ever since the dawn of electrical experimentation, it had been observed that such a device, whose ancestor is the Leyden jar, could store electrical charge. A condenser's capacity C, measured in *farads*, is defined as the ratio of the stored electrical charge Q_C, measured in coulombs, to the corresponding potential difference V_C between its electrodes, measured in volts. In other words,

$$C = \frac{Q_C}{V_C}.$$ (8.57)

Voltage and charge have the same sign. The dotted arrow representing voltage may therefore also represent charge. It is clear from the diagram that the current, as shown, corresponds to a depletion of charge per unit time. We shall therefore write

$$(Q_C)_{i+1} = (Q_C)_i - (I_C)_i \, \Delta t.$$

For the sake of simplicity, we shall rewrite the above expression as follows, where τ represents an *infinitesimal* time increment

$$Q_C' = Q_C - I_C \tau,$$

and we get, with $Q_C = CV_C$ and $Q_C' = CV_C'$,

Figure 8.15. The capacitor.

Figure 8.16. The inductor.

$$V_C' = V_C - \frac{I_C}{C}\tau.$$ (8.58)

The Inductor

The inductor simply consists of a coil of wire (fig. 8.16), and its inductance depends on the number of turns, as well as the magnetic properties (permeability) of the core around which it is wound.[4] High values of inductance are achieved with high permeability iron compound cores. Inductors that are found in high-frequency circuits have small inductance values and are wound on an air core. It was discovered that when a battery is connected across an inductor, the latter will resist the onset of current, and when the battery is disconnected, the inductor will resist the current's decay. With the directions chosen for the arrows in the diagram, inductance L is defined as the ratio of the voltage across the inductor, to the rate of current decay within that inductor. The equation corresponding to figure 8.16 is thus

$$L = \frac{V_L}{(I_L' - I_L)/\tau},$$

from which it follows that

$$I_L' = I_L + \frac{V_L}{L}\tau.$$ (8.59)

The Series RLC Circuit

Having defined the behavior of each individual component, we now turn our attention to the circuit in figure 8.17, in which the three components are connected *in series*. Let us ignore, for the time being,

[4] Any given length of wire possesses nonzero inductance.

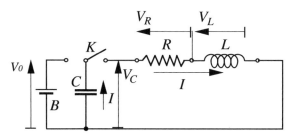

Figure 8.17. The series RLC circuit.

battery B on the left-hand side of the diagram and imagine that key K is closed toward the right-hand side. We can write

$$I_C = I_R = I_L = I, \qquad V_C = V_L + V_R,$$

from which it follows that

$$I' = I + \frac{V_L}{L}\tau = I + \frac{V_C - RI}{L}\tau.$$

Putting $V_C = V$, we get

$$I' = \frac{V}{L}\tau + I\left(1 - \frac{R}{L}\tau\right), \tag{8.60}$$

and equation (8.58) becomes

$$V' = V - \frac{I}{C}\tau. \tag{8.61}$$

Equations (8.60) and (8.61) can be combined in matrix form

$$\begin{bmatrix} V_{i+1} \\ I_{i+1} \end{bmatrix} = \begin{bmatrix} 1 & -\dfrac{1}{C}\tau \\ \dfrac{1}{L}\tau & 1 - \dfrac{R}{L}\tau \end{bmatrix} \begin{bmatrix} V_i \\ I_i \end{bmatrix} \tag{8.62}$$

where V is the voltage across the condenser, and I the current through the circuit. Putting

$$\omega = \sqrt{\frac{1}{LC} - \frac{R^2}{4L^2}}, \qquad (8.63)$$

ω is a real number when the circuit is *underdamped* (see the appendix). Substituting in equations (8.47) and (8.48) for very small values of time increment τ, we obtain, when τ is very small,

$$V = e^{-\frac{R}{2l}t}\left(V_0\left(\cos \omega t + \frac{R}{2L\omega}\sin \omega t \right) - I_0\frac{1}{C\omega}\sin \omega t \right), \qquad (8.64a)$$

$$I = e^{-\frac{R}{2l}t}\left(V_0\frac{1}{L\omega}\sin \omega t + I_0\left(\cos \omega t - \frac{R}{2L\omega}\sin \omega t \right) \right),$$

$$\text{with } \omega = \sqrt{\frac{1}{LC} - \frac{R^2}{4L^2}}. \qquad (8.45b)$$

Here $f = 2\pi/\omega$ is called the *resonance frequency* of the circuit, $L\omega$ is the *impedance* of inductance L, and $1/C\omega$ is that of capacitance C at frequency f. Ratio $L\omega/R$ is denoted Q and called the *figure of merit* of the circuit. When that figure is high, we get, when R/L is very small,

$$V = e^{-\frac{R}{2L}t}\left(V_0 \cos \omega t - I_0\sqrt{\frac{L}{C}}\sin \omega t \right), \qquad (8.65a)$$

$$I = e^{-\frac{R}{2L}t}\left(V_0\sqrt{\frac{C}{L}}\sin \omega t + I_0 \cos \omega t \right), \quad \text{with } \omega = \sqrt{\frac{1}{LC}}. \qquad (8.65b)$$

We can now perform the following physical experiment. At the outset, key K is open. Current $I = 0$ and voltage $V = V_C = 0$. Key K is lowered on the left-hand side, connecting the capacitor across battery B. The capacitor is fully charged, and the voltage across its terminals is V_0. Key K is now lowered on the right-hand side, marking time $t = 0$, with $V = V_0, I_0 = 0$. From that instant on, voltage V across the capacitor and current I flowing through the circuit are given by

$$V = V_0 e^{-\frac{R}{2L}t} \cos \omega t, \qquad I = V_0 e^{-\frac{R}{2L}t}\sqrt{\frac{C}{L}}\sin \omega t, \quad \text{with } \omega = \sqrt{\frac{1}{LC} - \frac{R^2}{4L^2}}. \qquad (8.66)$$

APPENDIX: FINITE DIFFERENCE EQUATIONS

Consider the *second-order finite difference equation*

$$X_{i+2} + a_1 X_{i+1} + a_2 X_i = 0,$$

where a_1 and a_2 are constants and $a_2 \neq 0$. This can be rewritten, for simplicity,

$$X'' + a_1 X' + a_2 X = 0.$$

The search for a solution of the form $X_n = m^n$ leads to the *characteristic equation*

$$m^2 + a_1 m + a_2 = 0,$$

whose roots are

$$m_1 = \frac{-a_1 + \sqrt{a_1^2 - 4a_2}}{2}, \qquad m_2 = \frac{-a_1 - \sqrt{a_1^2 - 4a_2}}{2}.$$

The calculus of finite differences teaches us the following:

1. If $a_1^2 > 4a_2$, the roots are real and different, and the solution is of the form

$$X_n = C_1 m_1^n + C_2 m_2^n.$$

   The system is said to be *overdamped*.

2. If $a_1^2 = 4a_2$, there exists a single real root m, and the solution is of the form

$$X_n = (C_1 + nC_2)m^n.$$

   The system is said to be *critically damped*.

3. If $a_1^2 < 4a_2$, the roots are complex and different. Putting $\sqrt{4a_2 - a_1^2} = \beta$, and $-a_1 = \alpha$, the roots are $m_1 = \alpha + i\beta$, $m_2 = \alpha - i\beta$, where $i = \sqrt{-1}$. The system is said to be underdamped, and its solution is of the form

$$X_k = Ar^k \cos(k\varphi + B), \quad \text{where } r = \sqrt{\alpha^2 + \beta^2}, \varphi = \tan^{-1}\frac{\beta}{\alpha},$$

$$r = \frac{\alpha}{\cos\varphi} = \sqrt{\alpha^2 + \beta^2}$$

A, B are determined by the *initial*, or *boundary*, conditions. Consider now the matrix equation

$$\begin{bmatrix} X_{i+1} \\ Y_{i+1} \end{bmatrix} = \begin{bmatrix} 1 & -b \\ d & c \end{bmatrix} \begin{bmatrix} X_i \\ Y_i \end{bmatrix}, \tag{8.67}$$

where b, c, d are positive numbers. This will be rewritten, for simplicity,

$$\begin{bmatrix} X' \\ Y' \end{bmatrix} = \begin{bmatrix} 1 & -b \\ d & c \end{bmatrix} \begin{bmatrix} X \\ Y \end{bmatrix},$$

and we write

$$X' = X - bY, \quad \text{hence } Y = \frac{X - X'}{b},$$

$$X'' = X' - bY',$$

and with $Y' = dX + cY$, we get

$$X'' = X' - b\left(dX + c\frac{X - X'}{b} \right) = X'(1 + c) - X(bd + c),$$

hence the second-order difference equation

$$X'' - X'(1 + c) + X(bd + c) = 0.$$

To that equation corresponds the second-degree *auxiliary*, or *characteristic*, equation

$$m^2 - m(1 + c) + (bd + c) = 0,$$

whose roots are

$$m = \frac{(1 + c) \pm \sqrt{(1 - c)^2 - 4bd}}{2}.$$

Let us now turn our attention to the matrix equation

$$\begin{bmatrix} X_{i+1} \\ Y_{i+1} \end{bmatrix} = \begin{bmatrix} 1 & -\gamma\tau \\ \lambda\tau & 1 - 2\rho\tau \end{bmatrix} \begin{bmatrix} X_i \\ Y_i \end{bmatrix}, \tag{8.68}$$

where γ, λ, ρ are real positive numbers and τ is also a positive number, which we will later refer to as the *elementary time increment*. Substituting in equation (8.68), the roots of the corresponding characteristic equation are

$$m = (1 - \rho\tau) \pm \sqrt{\rho^2 - \gamma\lambda}\tau.$$

The system is

> *overdamped if* $\rho^2 > \gamma\lambda$
> *critically damped if* $\rho^2 = \gamma\lambda$
> *underdamped if* $\rho^2 < \gamma\lambda$.

We shall only address the latter case, because of its relevance to the study of spirals. Putting $\omega = \sqrt{\gamma\lambda - \rho^2}$, which is referred to as the *angular velocity in the complex plane* (a mathematical artefact not to be confused with physical angular velocity of a rotating body) and sometimes *pulsation* in French, we get

$$m = (1 - \rho\tau) \pm i\omega\tau,$$

$$\varphi = \tan^{-1} \frac{\omega\tau}{1 - \rho\tau},$$

$$X_n = Ar^n \cos(n\varphi + B), \quad Y_n = Cr^n \cos(n\varphi + D).$$

Turning first to X_n, we get

$$X_n = A \left(\frac{1 - \rho\tau}{\cos\varphi} \right)^n (\cos n\varphi \cos B - \sin n\phi \sin B),$$

hence

$$X_1 = A \frac{1 - \rho\tau}{\cos\phi} \left(\frac{X_0}{A} \cos\varphi - \frac{\sqrt{A^2 - X_0^2}}{A} \sin\varphi \right)$$

$$= (1 - \rho\tau)(X_0 - \sqrt{A^2 - X_0^2} \tan\varphi).$$

Remembering that $X_1 = X_0 - \gamma\tau Y_0$, we can write

$$X_0 - \gamma\tau Y_0 = X_0 - \rho\tau X_0 - (1 - \rho\tau)\sqrt{A^2 - X_0^2} \tan\varphi,$$

$$\sqrt{A^2 - X_0^2} = \frac{\gamma Y_0 - \rho X_0}{\sqrt{\gamma\lambda - \rho^2}} = \frac{\gamma Y_0 - \rho X_0}{\omega},$$

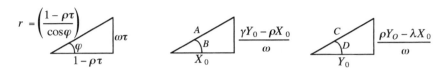

and we obtain

$$X_n = \left(\frac{1 - \rho\tau}{\cos\varphi}\right)^n \left(X_0 \cos n\varphi - \frac{\gamma Y_0 - \rho X_0}{\omega} \sin n\varphi\right),$$

or

$$X_n = \left(\frac{1 - \rho\tau}{\cos\varphi}\right)^n \left(X_0 \left(\cos n\varphi + \frac{\rho}{\omega} \sin n\varphi\right) - Y_0 \frac{\gamma}{\omega} \sin n\varphi\right).$$

We now turn our attention to Y_n. Following the above strategy, we eventually obtain

$$Y_n = \left(\frac{1 - \rho\tau}{\cos\varphi}\right)^n \left(X_0 \frac{\lambda}{\omega} \sin n\varphi + Y_0 \left(\cos n\varphi - \frac{\rho}{\omega} \sin n\varphi\right)\right).$$

Positional Number Systems

Denaria enum ex institute hominum, non ex necessitate
naturae ut vulgus arbitratur, et sane satis inepte, posita est.
(The decimal system has been established, somewhat
foolishly to be sure, according to man's customs, not from
a natural necessity as most people would think)
(*Blaise Pascal, De Numerus multiplicibus*)

This chapter is an introduction to the study of positional number
representation. It is an essential preamble to the following chapter,
which deals with fractals. The chapter opens with an analysis of the
division algorithm, upon which the system is predicated. Following a
brief introduction to codes and their algebraic representations, the
mixed base notion is introduced.

DIVISION

Given arbitrary numbers 24.5 and 7.2, respectively referred to as
the *dividend* and the *divisor*, we can write the following:

$$24.5 = \mathbf{1} \times 7.2 + 17.3,$$

$$24.5 = \mathbf{2} \times 7.2 + 10.1,$$

$$24.5 = \mathbf{3} \times 7.2 + 2.9,$$

$$24.5 < \mathbf{4} \times 7.2.$$

Inasmuch as 3 is the largest non-negative integer whose product by
7.2 does not exceed 24.5, it is called the *quotient*. The number 2.9 is
the *remainder, or residue*, and is, by construction, smaller than the
divisor. The above process, which leads to the determination of the
quotient and remainder, is the *division algorithm*.

Generally, given non-negative dividend D and positive divisor d, the division algorithm yields

$$D = dq + \rho \quad (0 \leq \rho < d, q \text{ is an integer}). \tag{9.1}$$

We can easily show that to the pair (D, d) corresponds one and only one pair of non-negative numbers satisfying equation (9.1), namely, the quotient q and remainder ρ.

Given integer $m > 1$, *integral variable* δ is said to belong to *range m* if it can take on integral values $0, 1, 2, \ldots m - 1$, and only these values. This is written

$$0 \leq \delta \leq m - 1. \tag{9.2}$$

(We can indifferently say that the variable *takes on, assumes,* or *is assigned* these values.) Let δ_0, δ_1, γ be integral variables respectively belonging to the ranges m_0, m_1, $m_0 m_1$, that is,

$$0 \leq \delta_0 \leq m_0 - 1, \quad 0 \leq \delta_1 \leq m_1 - 1, \quad 0 \leq \gamma \leq m_0 m_1 - 1. \tag{9.3}$$

Variables δ_0 and δ_1 are *independent*, meaning that each variable can assume any value within its range, irrespective of the value assumed by the other.

Given some agreed upon ordering (m_0, m_1) of ranges m_0, m_1, any particular ordered pair of values (δ_0, δ_1) assigned to variables δ_0, δ_1 within their respective ranges is called a *configuration*. There are obviously $m_0 \times m_1$ distinct configurations. For example, with $m_0 = 3$ and $m_1 = 2$, we have $0 \leq \delta_0 \leq 2$, $0 \leq \delta_1 \leq 1$, and $0 \leq \gamma \leq 5$.

A one-to-one correspondence, or *mapping*, can be arbitrarily defined between the configurations (δ_0, δ_1), which are $m_0 m_1$ in number, and the values assigned to variable γ, which are also $m_0 m_1$ in number. Each mapping is referred to as a *code*. Within a given code, any particular configuration (δ_0, δ_1) is said to be *representation* of the corresponding value of γ.

Table 9.1 shows two out of the $6! = 720$ different possible codes corresponding to $m_0 = 3$, $m_1 = 2$. If we regard γ as a *function* of independent variables δ_0, δ_1, we can write polynomial expressions of

TABLE 9.1

Linear Codes

Code	a		b	
γ	δ_1	δ_0	δ_1	δ_0
0	0	0	0	0
1	0	1	1	0
2	0	2	0	1
3	1	0	1	1
4	1	1	0	2
5	1	2	1	2

γ in terms of δ_0, δ_1 for any one out of the 720 codes. The polynomials corresponding to the codes of Table 9.1 are

$$\text{Code } a \qquad \gamma = \delta_0 + 3\delta_1,$$
$$\text{Code } b \qquad \gamma = \delta_1 + 2\delta_0.$$

Observe the simplicity of these first-degree polynomials, whose corresponding codes are said to be *linear*. Any code other than the above two will contain powers greater than 1 in any one of the variables, or products of the variables, such as $\delta_0\delta_1$. Returning to equation (9.1) and substituting integers γ for dividend D, m_0 for divisor d, and δ_0 for remainder ρ, we can write

$$\gamma = \delta_0 + m_0\,\delta_1, \quad \text{with} \quad 0 \leq \delta_0 \leq m_0 - 1. \tag{9.4}$$

Making $0 \leq \delta_1 \leq m_1 - 1$ implies $(0 \leq \gamma \leq m_0m_1 - 1)$, and conversely. Generally then, given ranges m_0, m_1 and variables δ_0, δ_1, γ satisfying statements (9.3), it follows that for every value of γ there exists one and only one configuration (δ_0, δ_1) satisfying equation (9.4), and, conversely, to every configuration (δ_0, δ_1) corresponds one and only one value of γ.

Moving on, consider ranges m_0, m_1, m_2 and independent variables δ_0, δ_1, δ_2 respectively belonging to these ranges. Consider also variable γ belonging to range $m_0m_1m_2$:

189

$$0 \le \gamma \le m_0 m_1 m_2 - 1. \tag{9.5}$$

The notion of *configuration* may now be extended to three integral variables. To each value of γ corresponds one and only one configuration $(\delta_0, \delta_1, \delta_2)$ satisfying equation (9.6), and conversely,

$$\gamma = \delta_0 + m_0(\delta_1 + m_1\,\delta_2) = \delta_0 + m_0\delta_1 + m_0 m_1 \delta_2 \tag{9.6}$$

or

$$\gamma = \pi_0\delta_0 + \pi_1\delta_1 + \pi_2\delta_2. \tag{9.7}$$

Coefficients $\pi_0 = 1$, $\pi_1 = m_0$, $\pi_2 = m_0 m_1$ are referred to as *weights*. Equation (9.7) defines a linear code and constitutes the foundation upon which *positional number systems* are built, such as the decimal and binary systems.

As an exercise, we shall assign integral values 2, 3, 4, to ranges m_0, m_1, m_2 in all $3! = 6$ possible ways, and display the polynomial corresponding to each assignment. These are

	m_2	m_1	m_0		$\gamma = \pi_0\delta_0 + \pi_1\delta_1 + \pi_2\delta_2$
Code 1	(4,	3,	2)	$\rightarrow$	$\gamma = \delta_0 + 2\delta_1 + 6\delta_2$
Code 2	(4,	2,	3)	$\rightarrow$	$\gamma = \delta_0 + 3\delta_1 + 6\delta_2$
Code 3	(3,	4,	2)	$\rightarrow$	$\gamma = \delta_0 + 2\delta_1 + 8\delta_2$
Code 4	(3,	2,	4)	$\rightarrow$	$\gamma = \delta_0 + 4\delta_1 + 8\delta_2$
Code 5	(2,	4,	3)	$\rightarrow$	$\gamma = \delta_0 + 3\delta_1 + 12\delta_2$
Code 6	(2,	3,	4)	$\rightarrow$	$\gamma = \delta_0 + 4\delta_1 + 12\delta_2$

For $\gamma = 14$, the following configurations correspond to codes 1–6:

	δ_2	δ_1	δ_0		
Code 1	(2,	1,	0)	$\rightarrow$	$14 = 0 + 2 \times 1 + 6 \times 2$
Code 2	(2,	0,	2)	$\rightarrow$	$14 = 2 + 3 \times 0 + 6 \times 2$
Code 3	(1,	3,	0)	$\rightarrow$	$14 = 0 + 2 \times 3 + 8 \times 1$

Code 4	(1,	1,	2)	$\rightarrow$	$14 = 2 + 4 \times 1 + 8 \times 1$
Code 5	(1,	0,	2)	$\rightarrow$	$14 = 2 + 3 \times 0 + 12 \times 1$
Code 6	(1,	0,	2)	$\rightarrow$	$14 = 2 + 4 \times 0 + 12 \times 1.$

Symbols m_0, m_1, m_2 and δ_0, δ_1, δ_2 are arranged in *Arabic order*, that is, from right to left.

MIXED BASE POSITIONAL SYSTEMS

Base b is defined as the ordered sequence of integers

$$b = (m_0, m_1, m_2, m_3, \ldots, m_i, \ldots, m_{n-1}), \quad m_i > 1 \text{ for all } i. \tag{9.8}$$

It is said to be of length n, and integers m_0, m_1, m_2, $\ldots$, m_i, $\ldots$, m_{n-1} are referred to as the *base ranges*, or *base radices* of base b. Any particular assignment of integral values to the variables $(\delta_0, \delta_1, \delta_2, \ldots, \delta_i, \ldots, \delta_{n-1})$, within their corresponding ranges, is referred to as a *conformable configuration*. In other words, a configuration is said to be *conformable* to base b if

$$0 \le \delta_i \le m_i - 1, \quad \text{for every } i. \tag{9.9}$$

To each index i is assigned *weight* π_i:

$$\pi_i = m_0 m_1 m_2 m_3, \ldots, m_{i-1}, \quad \text{with } \pi_0 = 1. \tag{9.10}$$

It follows that to any value of integer γ belonging to range π_n, (i.e., $0 \le \gamma \le \pi_n - 1$), corresponds one and only one conformable configuration $(\delta_0, \delta_1, \delta_2, \ldots, \delta_i, \ldots, \delta_{n-1})$ satisfying equation (9.11), and conversely;

$$\gamma = \delta_0 + m_0(\delta_1 + m_1(\delta_2 + m_2(\delta_3 + m_3(\delta_4 + \cdots + m_{n-2}\delta_{n-1}))))$$

$$= \pi_0 \delta_0 + \pi_1 \delta_1 + \pi_2 \delta_2 + \pi_3 \delta_3 + \cdots + \pi_{n-1}\delta_{n-1}. \tag{9.11}$$

The brackets are intended to provide a clue to the proof, which may be done by induction over n. By virtue of that one-to-one correspondence, we shall say that $(\delta_0, \delta_1, \delta_2, \ldots, \delta_i, \ldots, \delta_{n-1})$ is the *representation of γ, base b*. That representation will be written *from right to left*, consistent with Arabic numeration. A dot may be placed at the right-hand extremity, corresponding to $i = 0$. Integer δ_i will be referred to as the *ith digit of γ*, or digit of *rank i*, base b, and written $(\delta_i^\gamma)_b$ or simply $(\delta_i^\gamma$ when base b is clear from context. We shall write

$$\gamma = (\delta_{n-1}^\gamma, \ldots, \delta_i^\gamma, \ldots, \delta_2^\gamma, \delta_1^\gamma, \delta_0^\gamma.)_b = \sum_{i=0}^{n-1} \delta_i^\gamma \pi_i. \quad (9.12)$$

Referring to Table 9.2, we get

$$14 = (210.)_a = (202.)_b = (130.)_c = (112.)_d = (102.)_e = (102.)_g$$

with

$$a = (4, 3, 2.), b = (4, 2, 3.), c = (3, 4, 2.),$$
$$d = (3, 2, 4.), e = (2, 4, 3.), g = (2, 3, 4).$$

When $m_1 = m_2 = \ldots m_i \ldots = m$, we are in the presence of a *uniform base-m*, or *uniform radix-m* system, for which $\pi_i = m^i$. Thus,

$$\gamma = (\delta_{n-1}^\gamma, \ldots, \delta_i^\gamma, \ldots, \delta_2^\gamma, \delta_1^{\gamma rb}, \delta_0^\gamma.)_m = \sum_{i=0}^{n-1} \delta_i^\gamma m^i. \quad (9.13)$$

Otherwise, the system is referred to as *mixed base*, or *mixed radix*. Table 9.2 displays the correspondence between γ, which is shown base 10 in bold characters, and its representations base 2, 3, and 5. Base 10 is the familiar Hindu-Arabic decimal system, and base 2 is the no less familiar binary system. Innumerable other bases were proposed over the centuries, the most notable of which were the Mesopotamian sexagesimal and the Mayan vigesimal systems. Buffon advocated the use of the duodecimal system, whose base 12 is divisible by 2, 3, 4, and 6, a considerable virtue in his opinion, and duodecimal fanatics went as far as publishing tables of logarithms in that system. Other systems, though not positional, were also found to have particular merits, such as the binomial, the Fibonacci, and Shannon's balanced decimal system.

Unless the base is 10, or it is clear from the context, and whether that base is uniform or not, number representations will be surrounded

TABLE 9.2
Base-*m* Positional Numeration
for *m* = 2, 3, 5

decimal	base 5	base 3	base 2
0	0	0	0
1	1	1	1
2	2	2	10
3	3	10	11
4	4	11	100
5	10	12	101
6	11	20	110
7	12	21	111
8	13	22	1000
9	14	100	1001
10	20	101	1010
11	21	102	1011
12	22	110	1100
13	23	111	1101
14	24	112	1110
15	30	120	1111
16	31	121	10000
17	32	122	10001
18	33	200	10010
19	34	201	10011
20	40	202	10100
21	41	210	10101
22	42	211	10110
23	43	212	10111
24	44	220	11000

by brackets, followed by a symbol identifying the base. To alleviate inconsistencies, number representations as well as base range sequences will be written from right to left. Commas between integers may or may not be used. For example,

$$(101.)_2 = (12.)_3 = 5,$$

$$(211.)_a = 15, \quad \text{with } a = (4, 3, 2.), \qquad (132.)_b = 2, \quad \text{with } b = (3, 4, 2.).$$

Clearly, one may add any number of zeros to the left without changing the value of the number being represented. The leftmost nonzero digit is called the *highest significant digit,* and all digits to its right are significant. If s is the number of significant digits of integer γ, in other words if $s - 1$ is the rank of its highest significant digit, then

$$\pi_{s-1} \leq \gamma \leq \pi_s - 1 \quad \text{or} \quad \pi_{s-1} - 1 < \gamma < \pi_s. \qquad (9.14)$$

For example, any number γ with three significant decimal digits is such that

$$100 \leq \gamma \leq 999 \quad \text{or} \quad 99 < \gamma < 1000.$$

Similarly, any number γ with four significant binary digits, or *bits,* is such that

$$(1000.)_2 \leq \gamma \leq (1111.)_2 \quad \text{or} \quad (111.)_2 < \gamma < (10000.)_2$$

that is,

$$8 \leq \gamma \leq 15 \quad \text{or} \quad 7 < \gamma < 16.$$

Examples abound in literature where reference is made to "binary numbers" or "decimal numbers." Obviously, these attributes cannot be conferred upon the numbers themselves, but upon their representations. For the sake of simplicity, we shall nonetheless use linguistic shortcuts in extending to the number itself the attributes of its representation, and vice versa. Another shortcut has been to write $\gamma = (101.)_2 = (12.)_3 = 5$, that is, the integer whose decimal representation is 5 also has representations $(101.)_2$ in the binary, and $(12.)_3$ in the ternary, systems. All four representations, including symbol γ itself, correspond to the same integer, an abstract mathematical being, and are equivalent. We invite the reader to always maintain the intellectual distinction between a number and its representations.

FINDING THE DIGITS OF AN INTEGER

Given base b, and integer γ expressed in some different base, say, decimal, several algorithms allow the determination of the integer's digits, base b. The methods described below (as well as every other method for that matter) follow directly from the division algorithm and proceed according to the following steps.

Method 1

1. Divide γ by m_0 The quotient is q_1 and the remainder δ_0^γ
2. Divide q_1 by m_1 The quotient is q_2 and the remainder δ_1^γ
3. Divide q_2 by m_2 The quotient is q_3 and the remainder δ_2^γ

. . .

s. Divide q_{s-1} by m_{s-1} The quotient is 0 and the remainder δ_{s-1}

Examples with $b = (5, 4, 3, 2.)$: $\pi_0 = 1$, $\pi_1 = 2$, $\pi_2 = 6$, $\pi_3 = 24$, $\pi_4 = 120$.

1. $\gamma = 115 = 2 \times 57 + 1$
$57 = 3 \times 19 + 0$
$19 = 4 \times \;\; 4 + 3$
$4 = 5 \times \;\; 0 + 4$ $115 = (4301.)_b$

2. $\gamma = 24 = 2 \times 12 + 0$
$12 = 3 \times 4 + 0$
$4 = 4 \times 1 + 0$
$1 = 5 \times 0 + 1$ $24 = (1000.)_b = \pi_3$

3. $\gamma = 119 = 2 \times 59 + 1$
$59 = 3 \times 19 + 2$
$19 = 4 \times 4 \;\; + 3$
$4 = 5 \times 0 + 4$ $119 = (4321.)_b = \pi_4 - 1$
                          (119 is the largest conformable
                          integer of length 4).

Method 2. The base conversion algorithm described above may be reexpressed step by step as follows, where symbol [A] represents the integral part of number A.

1. $\gamma = m_0 \left[\dfrac{\gamma}{m_0} \right] + \delta_0^\gamma$

2. $\left[\dfrac{\gamma}{m_0} \right] = m_1 \left[\dfrac{\left[\dfrac{\gamma}{m_0} \right]}{m_1} \right] + \delta_1^\gamma = m_1 \left[\dfrac{\gamma}{m_0 m_1} \right] + \delta_1^\gamma$

3. $\left[\dfrac{\gamma}{m_0 m_1} \right] = m_2 \left[\dfrac{\gamma}{m_0 m_1 m_2} \right] + \delta_2^\gamma$

. . .

and in general,

$$\delta_i^\gamma = \left[\frac{\gamma}{\pi_i} \right] - m_i \left[\frac{\gamma}{\pi_{i+1}} \right] = \left[m_i \frac{\gamma}{\pi_{i+1}} \right] - m_i \left[\frac{\gamma}{\pi_{i+1}} \right]. \qquad (9.15)$$

A theorem of number theory states that $[mA] - m\,[A] = ([mA] \bmod m)$, where $(N \bmod m)$ denotes the remainder of the division of N by m. Thus,

$$\delta_i^\gamma = \left(\left[\frac{\gamma}{\pi_i} \right] \bmod m_i \right). \qquad (9.16)$$

Returning to the example of $\gamma = 115$,

$$\delta_0^\gamma = \left(\left[\frac{115}{1} \right] \bmod 2 \right) = 1, \qquad \delta_2^\gamma = \left(\left[\frac{115}{6} \right] \bmod 4 \right) = 3,$$

$$\delta_1^\gamma = \left(\left[\frac{115}{2} \right] \bmod 3 \right) = 0, \qquad \delta_3^\gamma = \left(\left[\frac{115}{24} \right] \bmod 5 \right) = 4.$$

Using that method to convert $\gamma = 315$ from decimal to binary,

$$\delta_0^\gamma = \left(\left[\frac{315}{1} \right] \bmod 2 \right) = 1, \qquad \delta_5^\gamma = \left(\left[\frac{315}{32} \right] \bmod 2 \right) = 1,$$

$$\delta_1^\gamma = \left(\left[\frac{315}{2} \right] \bmod 2 \right) = 1, \qquad \delta_6^\gamma = \left(\left[\frac{315}{64} \right] \bmod 2 \right) = 0,$$

$$\delta_2^\gamma = \left(\left[\frac{315}{4}\right] \bmod 2\right) = 0, \qquad \delta_7^\gamma = \left(\left[\frac{315}{128}\right] \bmod 2\right) = 0,$$

$$\delta_3^\gamma = \left(\left[\frac{315}{8}\right] \bmod 2\right) = 1, \qquad \delta_8^\gamma = \left(\left[\frac{315}{256}\right] \bmod 2\right) = 1.$$

$$\delta_4^\gamma = \left(\left[\frac{315}{16}\right] \bmod 2\right) = 1.$$

Since $315 < 2^i$ for all powers i higher than 8, δ_8^γ is the highest significant digit, and the required number is $(1\ 0\ 0\ 1\ 1\ 1\ 0\ 1\ 1.)_2$. The first base-conversion method is *iterative*. In order to calculate digit δ_i^γ, it requires the calculation of digits δ_j^γ for every index j from 0 to $i - 1$. The second method has the advantage of being based upon an explicit formulation of δ_i^γ, allowing the calculation of every digit on its own, *d'emblée*, inasmuch, of course, as the calculation of a number's integral part can be regarded as deriving from a truly explicit formulation, not requiring iteration, notwithstanding the essentially algorithmic nature of the division process itself.

Fractals

The existence of these patterns challenges us to study those
forms that Euclid leaves aside as being "formless," to
investigate the morphology of the "amorphous."
(*Benoît Mandelbrot*)[1]

Much has been written about fractals since the seminal work of
Mandelbrot and its ramifications into practically every realm of human
endeavor, from the so-called exact sciences to the softer human sciences.
Fractals are observed everywhere and have caused a great deal of excite-
ment among serious scientists as well as tinkerers. It is not my intention
to write one more course on fractals. Neither is it my intention to cover
every aspect of fractals, such as the fractal dimension or the Mandelbrot
set. Rather, I believe that the approach presented in this chapter is
somewhat unusual, as it derives directly from number-theoretical considerations. Many classical fractals are obtained, as well as some original ones.

Departing from the traditional practice, which consists in number-
ing the rows of a matrix from 1 to m and its columns from 1 to m',
we shall respectively number them from 0 to $m - 1$, and 0 to
$m' - 1$, as in figure 10.1. That simple though basic change authorizes
the performance of arithmetic operations on row and column indices,
and a redefinition of the Kronecker product of matrices in terms of the
arithmetic operations thus performed.[2] Beyond the two-dimensional
matrix, similar products may be defined for higher dimensional lattices,
again in terms of arithmetic operations on their indices.

THE KRONECKER PRODUCT REVISITED

Consider the $m \times m'$ matrix M of Figure 10.1, which has m rows
and m' columns.

---

[1] *The Fractal Geometry of Nature* (New York: W.H. Freeman, 1981), p. 1.

[2] As a young student, I never ceased to be baffled by the fact that the top-left
matrix element was always denoted $A_{1,1}$ instead of $A_{0,0}$. After all, I thought, it is zero,
not one, that is the first integer, if not the first *natural* number!

$$\begin{bmatrix} M_{0,0} & M_{0,1} & \cdots & M_{0,m'-1} \\ M_{1,0} & M_{1,1} & \cdots & M_{1,m'-1} \\ \cdots & \cdots & \cdots & \cdots \\ M_{m-1,0} & M_{m-1,1} & & M_{m-1,m'-1} \end{bmatrix}$$

Fig. 10.1. Matrix M.

We write

$$M = [M_{\mu,\mu'}], \quad \text{with } 0 \le \mu \le m-1, \quad 0 \le \mu' \le m'-1, \quad (10.1)$$

meaning that the element located at the intersection of row μ and column μ' is $M_{\mu,\mu'}$. Element $M_{0,0}$ is referred to as the matrix's *origin*. Consider now the $r \times r'$ matrix R

$$R = [R_{\rho,\rho'}], \quad \text{with } 0 \le \rho \le r-1, \quad 0 \le \rho' \le r'-1. \quad (10.2)$$

We define the *Kronecker product* $G = M \otimes R$ as the matrix $G = [G_{\gamma,\gamma'}]$, where

$$G_{\gamma,\gamma'} = M_{\mu,\mu'} \times R_{\rho,\rho'}, \quad \text{for } \gamma = \mu + m\rho, \quad \gamma' = \mu' + m'\rho'. \quad (10.3)$$

In chapter IX we learned that to any integer pair (μ, ρ) corresponds one and only one value of integer $\gamma = \mu + m\rho$, and conversely, where m, r are natural numbers $(1, 2, 3, \ldots)$, and

$$0 \le \mu \le m-1, \quad 0 \le \rho \le r-1, \quad 0 \le \gamma \le mr-1.$$

To any element pair $(M_{\mu,\mu'}, R_{\rho,\rho'})$ thus corresponds one and only one element $G_{\gamma,\gamma'} = G_{\mu+m\rho,\mu'+m'\rho'}$, and conversely. Matrix G thus consists of mr rows and $m'r'$ columns, and we can write

$$G_{\gamma,\gamma'} = (M \otimes R)_{(\mu+m\rho),(\mu'+m'\rho')} = M_{\mu,\mu'} \times R_{\rho,\rho'}. \quad (10.4)$$

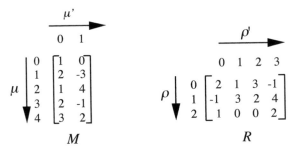

Fig. 10.2a. Matrices M and R.

Example. Figure 10.2a shows the 5 × 2 matrix M and the 3 × 4 matrix R. Figure 10.2b shows the 15 × 8 matrix $G = M \otimes R$. The left-hand margin shows the number $\gamma = 0$ to 14, along with its digits, base (3, 5.), namely, $0 \leq \mu \leq 4$ and $0 \leq \rho \leq 2$. Similarly, the top margin shows the number $\gamma' = 0$ to 7, along with its digits, base (4, 2.), namely, $0 \leq \mu' \leq 1$ and $0 \leq \rho' \leq 3$.

γ'			0	1	2	3	4	5	6	7
ρ'			0	0	1	1	2	2	3	3
μ'			0	1	0	1	0	1	0	1
γ	ρ	μ								
0	0	0	2	0	1	0	3	0	-1	0
1	0	1	4	-6	2	-3	6	-9	-2	3
2	0	2	2	8	1	4	3	12	-1	-4
3	0	3	4	-2	2	-1	6	-3	-2	1
4	0	4	6	4	3	2	9	6	-3	-2
5	1	0	-1	0	3	0	2	0	4	0
6	1	1	-2	3	6	-9	4	-6	8	-12
7	1	2	-1	-4	3	12	2	8	4	16
8	1	3	-2	1	6	-3	4	-2	8	-4
9	1	4	-3	-2	9	6	6	4	12	8
10	2	0	1	0	0	0	0	0	2	0
11	2	1	2	-3	0	0	0	0	4	-6
12	2	2	1	4	0	0	0	0	2	8
13	2	3	2	-1	0	0	0	0	4	-2
14	2	4	3	2	0	0	0	0	6	4

Fig. 10.2b. Matrix $G = M \otimes R$.

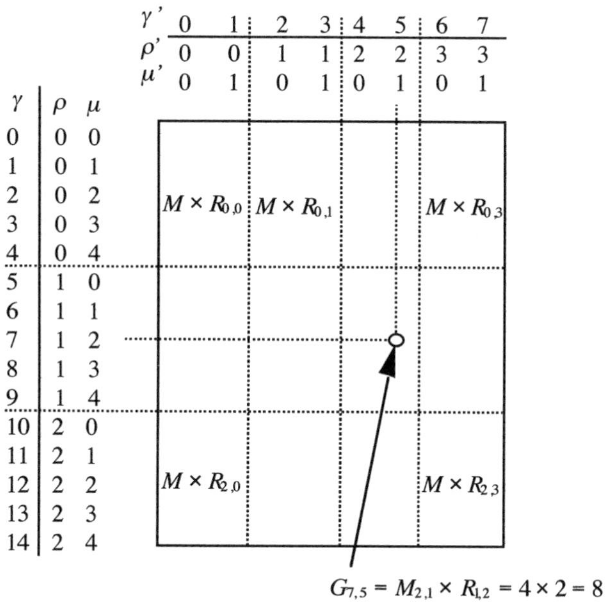

$$
\begin{array}{c|cccccccc}
\gamma' & 0 & 1 & 2 & 3 & 4 & 5 & 6 & 7 \\
\rho' & 0 & 0 & 1 & 1 & 2 & 2 & 3 & 3 \\
\mu' & 0 & 1 & 0 & 1 & 0 & 1 & 0 & 1
\end{array}
$$

γ	ρ	μ								
0	0	0								
1	0	1								
2	0	2	$M \times R_{0,0}$	$M \times R_{0,1}$				$M \times R_{0,3}$		
3	0	3								
4	0	4								
5	1	0								
6	1	1								
7	1	2								
8	1	3								
9	1	4								
10	2	0								
11	2	1								
12	2	2	$M \times R_{2,0}$					$M \times R_{2,3}$		
13	2	3								
14	2	4								

$$G_{7,5} = M_{2,1} \times R_{1,2} = 4 \times 2 = 8$$

Fig. 10.2c. Structure of matrix $G = M \otimes R$.

Figure 10.2c illustrates that in order to calculate the element of G that is located upon row γ and column γ', the number γ is written base $(r, m.)$, thus $(\rho, \mu.)$. Similarly, the number γ' is written base $(r', m'.)$, thus $(\rho', \mu'.)$. Having obtained the digits μ, μ', ρ, ρ', the elements $M_{\mu,\mu'}$ and $R_{\rho,\rho'}$ are sought from their respective matrices, and the required element is $G_{\gamma,\gamma'} = M_{\mu,\mu'} \times R_{\rho,\rho'}$. The macroscopic view of matrix G, shown in figure 10.2d, reveals that it consists of $r \times r'$ smaller matrices, each of which is obtained by multiplying matrix M by the element of matrix R that corresponds to the same geographic location within R as that of the smaller matrix within G. For that reason, M shall be called the *seed* matrix, and R the *template* matrix.

Associativity of the Kronecker Product

Given matrices $M = [M_{\mu,\mu'}]$, $R = [R_{\rho,\rho'}]$, $S = [S_{\sigma,\sigma'}]$, we can write

$$((M \otimes R) \otimes S)_{(\mu+m\rho)+mr\sigma, (\mu'+m'\rho')+m'r'\sigma'} = (M_{\mu,\mu'} \times R_{\rho,\rho'}) \times S_{\sigma,\sigma'}. \quad (10.5a)$$

Had we begun by forming the product $R \otimes S$, we could have written

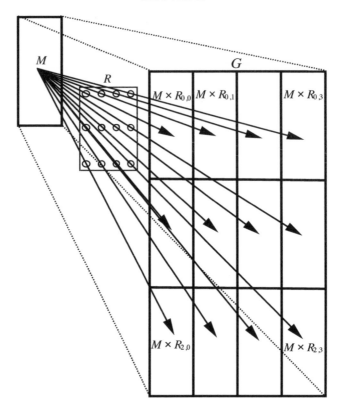

Fig. 10.2d. Macroscopic view of $G = M \otimes R$.

$$(M \otimes (R \otimes S))_{\mu+m(\rho+r\sigma),\,\mu'+m'(\rho'+r'\sigma')} = M_{\mu,\mu'} \times (R_{\rho,\rho'} \times S_{\sigma,\sigma'}).\quad (10.5b)$$

Putting

$$\gamma = \mu + m\rho + mr\sigma \quad \text{and} \quad \gamma' = \mu' + m'\rho' + m'r'\sigma',\quad (10.6)$$

we get

$$((M \otimes R) \otimes S)_{\gamma,\gamma'} = (M \otimes (R \otimes S))_{\gamma,\gamma'} = M_{\mu,\mu'} \times R_{\rho,\rho'} \times S_{\sigma,\sigma'}.\quad (10.7)$$

The Kronecker product is *associative*. We can thus delete the brackets and write

$$(M \otimes R) \otimes S = M \otimes (R \otimes S) = M \otimes R \otimes S.\quad (10.8)$$

The following two statements therefore imply each other:

$$
\begin{array}{cccc}
\mu' & 0 & 1 & 2 \\
\end{array}
\qquad
\begin{array}{ccc}
\rho' & 0 & 1 \\
\end{array}
$$

$$
\mu \quad
\begin{array}{c}
0 \\ 1 \\ 2
\end{array}
\begin{bmatrix}
1 & 0 & 0 \\
1 & 2 & 4 \\
1 & 3 & 9
\end{bmatrix}
\qquad
\rho \quad
\begin{array}{c}
0 \\ 1 \\ 2
\end{array}
\begin{bmatrix}
1 & 0 \\
2 & 3 \\
-2 & 1
\end{bmatrix}
\qquad
\begin{array}{l}
\sigma' \to \quad 0 \ 1 \\
\sigma = 0 \ [3 \ 4]
\end{array}
$$

$$
\qquad\qquad M \qquad\qquad\qquad\qquad R \qquad\qquad\qquad\qquad S
$$

Fig. 10.3a. Matrices M, R, S.

γ'	0	1	2	3	4	5	6	7	8	9	10	11
σ_1	0	0	0	0	0	0	1	1	1	1	1	1
ρ_1	0	0	0	1	1	1	0	0	0	1	1	1
μ_1	0	1	2	0	1	2	0	1	2	0	1	2

σ_0	ρ_0	μ_0												
0	0	0	3	0	0	0	0	0	4	0	0	0	0	0
0	0	1	3	6	12	0	0	0	4	8	16	0	0	0
0	0	2	3	9	27	0	0	0	4	12	36	0	0	0
0	1	0	6	0	0	9	0	0	8	0	0	12	0	0
0	1	1	6	12	24	9	18	36	8	16	32	12	24	48
0	1	2	6	18	54	9	27	81	8	24	72	12	36	108
0	2	0	-6	0	0	3	0	0	-8	0	0	4	0	0
0	2	1	-6	-12	-24	3	6	12	-8	-16	-32	4	8	16
0	2	2	-6	-18	-54	3	9	27	-8	-24	-72	4	12	36

Fig. 10.3b. Matrix $G = M \otimes R \otimes S$.

$$
\begin{aligned}
G_{2,1} &= M_{2,1} \times R_{0,0} \times S_{0,0} = 3 \times 1 \times 3 = 9 \\
G_{4,5} &= M_{1,2} \times R_{1,1} \times S_{0,0} = 4 \times 3 \times 3 = 36 \\
G_{8,7} &= M_{2,1} \times R_{2,0} \times S_{0,1} = 3 \times -2 \times 4 = -24 \\
G_{3,9} &= M_{0,0} \times R_{1,1} \times S_{0,1} = 1 \times 3 \times 4 = 12
\end{aligned}
$$

Fig. 10.3c. Elements of matrix G.

$$G = M \otimes R \otimes S \Leftrightarrow G_{\gamma,\gamma'} = M_{\mu,\mu'} \times R_{\rho,\rho'} \times S_{\sigma,\sigma'} \qquad (10.9a)$$

with

$$\gamma = \mu + m\rho + mr\sigma, \quad 0 \leq \gamma \leq mrs - 1 \qquad (10.9b)$$

and

$$\gamma' = \mu' + m'\rho' + m'r'\sigma', \quad 0 \leq \gamma' \leq m'r's' - 1 \qquad (10.9c)$$

(see figs. 10.3a and b).

Example. If symbol δ_i^{γ} denotes the ith digit of γ, base $(s, r, m.)$, and $\delta_i^{\gamma'}$ is the ith digit of γ', base $(s', r', m'.)$, in other words,

$$\gamma = (\delta_2^{\gamma}, \delta_1^{\gamma}, \delta_0^{\gamma}.)\text{base}(s,r,m.), \qquad \gamma' = (\delta_2^{\gamma'}, \delta_1^{\gamma'}, \delta_0^{\gamma'}.)\text{base}(s',r',m'.),$$
$$(10.10)$$

equation (10.9a) can be rewritten

$$(M \otimes R \otimes S)_{\gamma,\gamma'} = M_{\delta_0^{\gamma},\delta_0^{\gamma'}} \times R_{\delta_1^{\gamma},\delta_1^{\gamma'}} \times S_{\delta_2^{\gamma},\delta_2^{\gamma'}}. \qquad (10.11)$$

We can now generalize the above statements to any number of matrices, such as

$$M_{(0)} = \left[M_{(0)\mu_0,\mu_0'} \right], \quad M_{(1)} = \left[M_{(1)\mu_1,\mu_1'} \right], \quad M_{(2)} = \left[M_{(2)\mu_2,\mu_2'} \right], \ldots,$$

where

$$0 \leq \mu_0 < m_0 - 1, \qquad 0 \leq \mu_1 < m_1 - 1, \qquad 0 \leq \mu_2 < m_2 - 1, \ldots,$$
$$0 \leq \mu_0' < m_0' - 1, \qquad 0 \leq \mu_1' < m_1' - 1, \qquad 0 \leq \mu_2' < m_2' - 1, \ldots,$$

and write

$$(M_{(0)} \otimes M_{(1)} \otimes M_{(2)} \otimes \cdots)_{\gamma,\gamma'} = M_{(0)\delta_0^{\gamma},\delta_0^{\gamma'}} \times M_{(1)\delta_1^{\gamma},\delta_1^{\gamma'}} \times M_{(2)\delta_2^{\gamma},\delta_2^{\gamma'}} \times \cdots$$
$$(10.12a)$$

with

$$\gamma = (\ldots, \delta_2^{\gamma}, \delta_1^{\gamma}, \delta_0^{\gamma}.)\text{base}(\ldots m_2, m_1, m_0.),$$
$$\gamma' = (\ldots, \delta_2^{\gamma'}, \delta_1^{\gamma'}, \delta_0^{\gamma'}.)\text{base}(\ldots m_2', m_1', m_0'.). \qquad (10.12b)$$

Introducing symbol $\overset{n-1}{\underset{i=0}{\otimes}}$ to denote the repeated Kronecker product, we get

$$G = \overset{n-1}{\underset{i=0}{\otimes}} M_{(i)} \Leftrightarrow G_{\gamma,\gamma'} = \prod_{i=0}^{n-1} M_{(i)\delta_i^\gamma, \delta_i^{\gamma'}}. \qquad (10.13)$$

Once again, it must be stressed that δ_i^γ is the ith digit of γ, base $(m_{n-1}, \ldots, m_i, \ldots, m_2, m_1, m_0.)$, and δ_i^γ is the ith digit of γ', base $(m_{n-1}', \ldots, m_i', \ldots, m_2', m_1', m_0'.)$.

Matrix Order

We shall write, by way of definition, $M^{(1)} = M$, $M^{(2)} = M \otimes M$, $M^{(3)} = M \otimes M \otimes M$, and so on. In other words, the successive templates are identical to the seed. By virtue of the associativity of the Kronecker product, we can write

$$M^{(3)} = M \otimes M \otimes M = (M \otimes M) \otimes M = M \otimes (M \otimes M)$$
$$= M^{(2)} \otimes M = M \otimes M^{(2)}.$$

The statement can be easily generalized as

$$M^{(a)} \otimes M^{(b)} = M^{(b)} \otimes M^{(a)} = M^{(a+b)}, \quad a, b = 0, 1, 2, \ldots, \qquad (10.14)$$

and in particular,

$$M^{(n+1)} = M^{(n)} \otimes M = M \otimes M^{(n)}, \quad n = 0, 1, 2, \ldots. \qquad (10.15)$$

We shall deliberately put $M^{(0)} = U = [1]$, which statement is consistent with equation (10.14), and refer to U as the *unit matrix*.

With the above definitions, equation (10.13) becomes

$$M_{\gamma,\gamma'}^{(n)} = \prod_{i=0}^{n-1} M_{\delta_i^\gamma, \delta_i^{\gamma'}}, \quad n = 1, 2, \ldots, \qquad (10.16)$$

where

$$\delta_i^\gamma = (\delta_i^\gamma)_m, \qquad \delta_i^{\gamma'} = (\delta_i^{\gamma'})_{m'}. \qquad (10.17)$$

Matrix $M^{(n)}$ will be referred to as the matrix M of order n with respect to the Kronecker product. Figure 10.4 is an example of first and second order matrices.

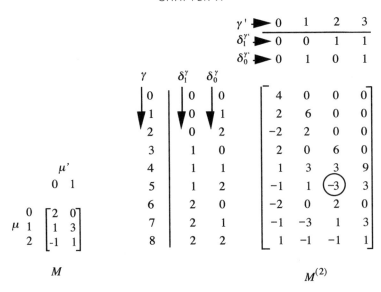

Fig. 10.4. First- and second-order matrices.

Commutativity of the Kronecker Product

In order to study the property of commutativity as it applies to the Kronecker product, we shall begin with figure 10.5, where, given matrices M and R, the products $G = M \otimes R$ and $H = R \otimes M$ are effected. The reader will have no difficulty understanding why the elements of both matrices are identical, albeit arranged differently. For these two matrices $m = 3$, $r = 3$, $m' = 3$, $r' = 2$, and, for all values of μ, μ', ρ, ρ' within their respective ranges, we have

$$G_{\mu+3\rho,\mu'+3\rho'} = H_{\rho+3\mu,\rho'+2\mu'} = M_{\mu,\mu'} \times R_{\rho,\rho'}.$$

If we choose $\mu = 2$, $\mu' = 2$, $\rho = 1$, $\rho' = 2$, we get

$$G_{2+3\times1,2+3\times0} = H_{1+3\times2,0+2\times2} = M_{2,2} \times R_{1,0},$$

$$G_{5,2} = H_{7,4} = 9 \times 2 = 18.$$

Elements $G_{5,2}$ and $H_{7,4}$ are circled within their respective matrices.

The Kronecker product of two matrices is thus generally noncommutative, as it shuffles their elements in different manners, depending on the order in which it is effected. However, we have seen, equation (10.14), that

$$M^{(a)} \otimes M^{(b)} = M^{(b)} \otimes M^{(a)} = M^{(a+b)}, \quad a, b = 0, 1, 2, \ldots.$$

Matrix M (rows μ_0, columns μ_1):

μ_1	0	1	2
μ_0			
0	1	0	0
1	1	2	4
2	1	3	9

Matrix R (rows ρ_0, columns ρ_1):

ρ_1	0	1
ρ_0		
0	1	0
1	2	3
2	-2	1

Fig. 10.5a. Matrices M and R.

γ'			0	1	2	3	4	5
ρ_1			0	0	0	1	1	1
μ_1			0	1	2	0	1	2
γ	ρ_0	μ_0						
0	0	0	1	0	0	0	0	0
1	0	1	1	2	4	0	0	0
2	0	2	1	3	9	0	0	0
3	1	0	2	0	0	3	0	0
4	1	1	2	4	8	3	6	12
5	1	2	2	6	(18)	3	9	27
6	2	0	-2	0	0	1	0	0
7	2	1	-2	-4	-8	1	2	4
8	2	2	-2	-6	-18	1	3	9

Fig. 10.5b. Product $G = M \otimes R$.

γ'			0	1	2	3	4	5
μ_1			0	0	0	1	1	1
ρ_1			0	1	2	0	1	2
η	μ_0	ρ_0						
0	0	0	1	0	0	0	0	0
1	0	1	2	3	0	0	0	0
2	0	2	-2	1	0	0	0	0
3	1	0	1	0	2	0	4	0
4	1	1	2	3	4	6	8	12
5	1	2	-2	1	-4	2	-8	4
6	2	0	1	0	3	0	9	0
7	2	1	2	3	6	9	(18)	27
8	2	2	-2	1	-6	3	-18	9

Fig. 10.5c. Producer $H = R \otimes M$.

That constitutes one instance where the Kronecker product is commutative. (Obviously, the commutativity of the Kronecker product of any matrix and the unit matrix is a particular case of the above, where either a or b is zero.)

Vectors

A vector is a unidimensional matrix, as in the following examples.

Examples

1. $[1\ 3\ 2\ 5] \otimes [3\ 4\ 2] = [3\ 9\ 6\ 15\ |\ 4\ 12\ 8\ 20\ |\ 2\ 6\ 4\ 10]$

2. $[3\ 4\ 2] \otimes [1\ 3\ 2\ 5] = [3\ 4\ 2\ |\ 9\ 12\ 6\ |\ 6\ 8\ 4\ |\ 15\ 20\ 10]$

3. Bottom row of matrix (10.5a):

$$[1\ 3\ 9] \otimes [-2\ 1] = [-2\ -6\ -18\ |\ 1\ 3\ 9]$$

4. Bottom row of matrix (10.5b):

$$[-2\ 1] \otimes [1\ 3\ 9] = [-2\ 1\ |\ -6\ 3\ |\ -18\ 9]$$

5. $\vec{V} = [2\ 1\ 3]$,
    $\vec{V}^{(2)} = [4\ 2\ 6\ |\ 2\ 1\ 3\ |\ 6\ 3\ 9]$,
    $\vec{V}^{(3)} = [8\ 4\ 12\ 4\ 2\ 6\ 12\ 6\ 18\ |\ 4\ 2\ 6\ 2\ 1\ 3\ 6\ 3\ 9\ |\ 12\ 6\ 18\ 6\ 3$
    $9\ 18\ 9\ 27]$.

In two-dimensional space, a vector can be regarded as either a one-row or a one-column matrix, and we may encounter the following situation

$$\underbrace{\begin{bmatrix} 1 \\ 3 \\ 2 \\ 5 \end{bmatrix}}_{M} \otimes \underset{R}{[3\ 4\ 2]} = \underset{R}{[3\ 4\ 2]} \otimes \underbrace{\begin{bmatrix} 1 \\ 3 \\ 2 \\ 5 \end{bmatrix}}_{M} = \begin{bmatrix} 3 & 4 & 2 \\ 9 & 12 & 6 \\ 6 & 8 & 4 \\ 15 & 20 & 10 \end{bmatrix}.$$

Here, $m' = 1$, hence $\mu' = 0$, and $r = 1$, hence $\rho = 0$. For any pair (γ, γ'), $0 \leq \gamma \leq m - 1$, $0 \leq \gamma' \leq r' - 1$, we thus have

$$(M \otimes R)_{\mu + m\rho,\, \mu' + m'\rho'} = (M \otimes R)_{\mu,\rho'} = M_{\mu,0} \times R_{0,\rho'},$$
$$(R \otimes M)_{\rho + r\mu,\, \rho' + r'\mu'} = (R \otimes M)_{\mu,\rho'} = R_{0,\rho'} \times M_{\mu,0},$$

hence

$$(M \otimes R) = (R \otimes M).$$

The Kronecker product of two vectors is therefore commutative when they belong to two different dimensions (as well as when one of the two vectors is the unit lattice, which constitutes a particular case of that configuration).

Fractal Lattices

Matrix $M^{(n)}$ of order $n > 1$ is said to be *gnomonic*, or *fractal*, when its *origin* $M_{0,0}$ is equal to 1. We shall later generalize the notion of Kronecker product to higher dimensional space, as well as to operators other than arithmetic multiplication.

Example. The adjective *fractal* is applied to $M^{(2)}$ in figure 10.6, by virtue of the character of *self-similarity* of that matrix, where one finds the matrix M embedded, unchanged, in the top left corner of $M^{(2)}$. For any value of n, the matrix $M^{(n)}$ thus finds itself embedded in the top left corner of $M^{(n+1)}$. That character of self-similarlity is high-lighted by figure 10.7. Similarly, vector $\vec{V}^{(n)}$ of order n is said to be fractal when its origin $\vec{V}_0 = 1$.

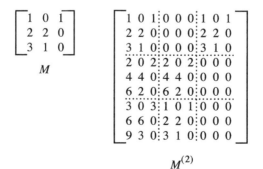

Fig. 10.6. Fractal matrix $M^{(2)}$.

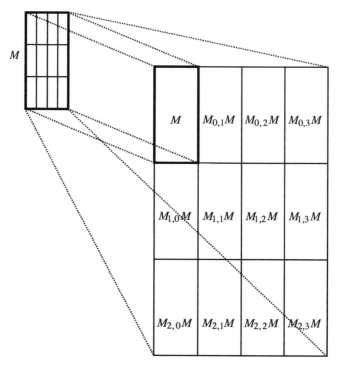

Fig. 10.7. Self-similarity of the fractal matrix.

Example. Given

$$\vec{V} = [1\,2\,3],$$
$$\vec{V}^{(2)} = [1\,2\,3\,|\,2\,4\,6\,|\,3\,6\,9],$$
$$\vec{V}^{(3)} = [1\,2\,3\,2\,4\,6\,3\,6\,9\,|\,2\,4\,6\,4\,8\,12\,6\,12\,18\,|\,3\,6\,9\,6\,12\,18\,9\,18\,27].$$

It follows from the definition of a gnomonic matrix $M^{(n)}$, $n > 1$, that

$$M^{(n)}_{\mu,\mu'} = M_{\mu,\mu'}, \quad \text{with } M_{0,0} = 1. \tag{10.18}$$

We therefore write

$$M^{(n)}_{\gamma,\gamma'} = \prod_{i=0}^{n-1} M^{(n)}_{\delta_i^\gamma, \delta_i^{\gamma'}}. \tag{10.19}$$

Hence the following definition of a fractal or gnomonic matrix: Matrix

$F = [F_{\gamma,\gamma'}]$ is said to be fractal, or gnomonic, if and only if integers m, m' can be found, such that for every element $F_{\gamma,\gamma'}$,

$$F_{\gamma,\gamma'} = \prod_i F_{\delta_i^\gamma, \delta_i^\gamma}, \quad i = 0, 1, 2, \ldots, \qquad (10.20a)$$

where

$$\gamma = (\ldots, \delta_i^\gamma, \ldots, \delta_2^\gamma, \delta_1^\gamma, \delta_0^\gamma.)_m, \quad \gamma' = (\ldots, \delta_i^{\gamma'}, \ldots, \delta_2^{\gamma'}, \delta_1^{\gamma'}, \delta_0^{\gamma'}.)_{m'}. \qquad (10.20b)$$

The omission of an upper limit above the repeated product symbol indicates that the range of the product exhausts all of the significant digits of both γ and γ'.

The growth pattern of a matrix from one order to the next is reminiscent of certain forms of animal and plant growth, and one cannot resist quoting D'Arcy Thompson one more time, when he writes "But the horn, or the snail-shell, is curiously different; for in these the presently existing structure is, so to speak, partly old and partly new. It has been conformed by successive and continuous increments; and each successive stage of growth, starting from the origin, remains as an *integral and unchanging* portion of the growing structure . . . . The elephant's tusk, the beaver's tooth, the cat's claw or the canary-bird . . . all alike consist of stuff secreted or deposited by living material; and in all alike the parts once formed remain in being, and are thenceforward incapable of change."

Pascal's Triangle and Lucas's Theorem

Having mentioned Lucas, I may recall that he was another of
the great amateurs, in the sense that, although he was
conversant with much of the higher mathematics of his day,
he refrained from working in the fashionable things of his
time in order to give his instinct for arithmetic free play.
(*Eric Temple Bell*)[3]

The French mathematician Blaise Pascal is perhaps best known for the triangle that bears his name, which provides an easy recursive method for the calculation of the number of combinations of a objects taken

[3] *The World of Mathematics*, ed. James R. Newman (New York: Simon & Schuster, 1956), p. 504.

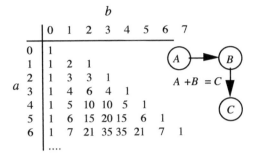

	b							
	0	1	2	3	4	5	6	7
0	1							
1	1	2	1					
2	1	3	3	1				
3	1	4	6	4	1			
4	1	5	10	10	5	1		
5	1	6	15	20	15	6	1	
6	1	7	21	35	35	21	7	1
	....							

Fig. 10.8a. Pascal's triangle and its construction.

b at a time. That number, which is referred to as the *binomial coefficient* and represented by the symbol $\binom{a}{b}$, is equal to $a!/b!(a - b)!$, where $a! = a \times (a - 1) \times (a - 2) \times \cdots \times 1$, and $0! = 1$. It is central to Newton's binomial expansion theorem, hence its name:

$$(a + b)^n = \binom{n}{0} a^n b^0 + \binom{n}{1} a^{n-1}b^1 + \binom{n}{2} a^{n-2}b^2 + \cdots + \binom{n}{i} a^{n-i}b^i$$

$$+ \cdots + \binom{n}{n} a^0 b^n, \quad \text{or} \quad (a + b) = \sum_{i=0}^{n} \binom{n}{i} a^{n-i}b^i.$$

$$(10.21)$$

Figure 10.8a shows how the triangle is constructed; the empty entries in the matrix are occupied by zeros.

Edouard Lucas (1842–1891) is another French mathematician who published a large number of theorems in number theory. He made substantial contributions to the study of Fibonacci numbers (it was Lucas who thus christened the famous sequence), which he notably used to prove that the thirty-nine-digit number $2^{127} - 1$ is prime. His *Récréations mathématiques*, a book of mathematical diversions, was published in 1893.[4] Lucas is the author of a beautiful little-known

---

[4] See Bell, *The World of Mathematics*: "Lucas's *Théorie des nombres, première partie*, 1891 (all issued, unfortunately), is a fascinating book for amateurs and the less academic professionals in the theory of numbers. His widely scattered writings should be collected, and his unpublished manuscripts sifted and edited" (p. 504).

```
1
1 1
1   1
1 1 1 1
1       1
1 1     1 1
1   1   1   1
1 1 1 1 1 1 1 1
1               1
1 1             1 1
1   1           1   1
1 1 1 1         1 1 1 1
1       1       1       1
1 1     1 1     1 1     1 1
1   1   1   1   1   1   1   1
1 1 1 1 1 1 1 1 1 1 1 1 1 1 1 1
```

Fig. 10.8b. $p = 2$.

```
1
1 1
1 2 1
1       1
1 1   1 1
1 2 1 1 2 1
1     2     1
1 1   2 2   1 1
1 2 1 2 1 2 1 2 1
```

Fig. 10.8c. $p = 3$.

```
1
1 1
1 2 1
1 3 3 1
1 4 6 4 1
1 5 3 3 5 1
1 6 1 6 1 6 1
1           1
1 1         1 1
1 2 1       1 2 1
1 3 3 1     1 3 3 1
1 4 6 4 1   1 4 6 4 1
1 5 3 3 5 1 1 5 3 3 5 1
1 6 1 6 1 6 1 1 6 1 6 1 6 1
1         2         1
1 1       2 2       1 1
1 2 1     2 4 2     1 2 1
1 3 3 1   2 6 6 2   1 3 3 1
1 4 6 4 1 2 1 5 1 2 1 4 6 4 1
1 5 3 3 5 1 2 3 6 6 3 2 1 5 3 3 5 1
1 6 1 6 1 6 1 2 5 2 5 2 5 2 1 6 1 6 1 6 1
```

Fig. 10.8e. $p = 7$.

```
1
1 1
1 2 1
1 3 3 1
1 4 1 4 1
1           1
1 1         1 1
1 2 1       1 2 1
1 3 3 1     1 3 3 1
1 4 1 4 1   1 4 1 4 1
1         2         1
1 1       2 2       1 1
1 2 1     2 4 2     1 2 1
1 3 3 1   2 1 1 2   1 3 3 1
1 4 1 4 1 2 3 2 3 2 1 4 1 4 1
```

Fig. 10.8d. $p = 5$.

theorem, which states that if the binomial coefficient $\binom{\gamma}{\gamma'}$ is written base p, where p is prime, we get

$$\binom{\gamma}{\gamma'} \equiv \prod_i \binom{\delta_i^\gamma}{\delta_i^{\gamma'}} \pmod p, \quad i = 0, 1, 2, \ldots, \quad \text{where}$$

$$\delta_i^\gamma = (\delta_i^\gamma)_p, \qquad \delta_i^{\gamma'} = (\delta_i^{\gamma'})_p.$$

213

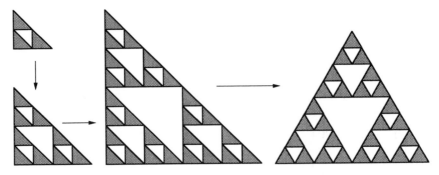

Fig. 10.9. Pascal's triangle modulo 2.

If the table of binomial coefficients $\begin{pmatrix} \gamma \\ \gamma' \end{pmatrix}$ is regarded as a square matrix $P = [p_{\gamma,\gamma'}]$, equation (10.22a) becomes

$$P_{\gamma,\gamma'} \equiv \prod_i P_{\delta_i^\gamma, \delta_i^{\gamma'}} \pmod{p}, \quad i = 0, 1, 2, \ldots, \qquad (10.22b)$$

signaling that $(P \bmod p)$ is a fractal matrix. Figure 10.9 shows four examples that illustrate that theorem, for $p = 2, 3, 5, 7$. If we let symbol P_q denote the top-left corner of matrix P comprising q rows and q columns, and symbol $P_q^{(n)}$ the nth order of that matrix, figure 10.8 illustrates the following statements:

Figure 10.8b: $P_{16}^{(1)} \equiv P_4^{(2)} \equiv P_2^{(4)} \pmod 2$,

Figure 10.8c: $P_9^{(1)} \equiv P_3^{(2)} \pmod 3$,

Figure 10.8d is a fragment of $P_5^{(n)} \pmod 5$,

Figure 10.8e is a fragment of $P_7^{(n)} \pmod 7$.

The theorem of Lucas can therefore be reformulated in the language of Kronecker products, as follows:

$$P_{p^a}^{(b)} \equiv P_{p^c}^{(d)} \pmod p,$$
$$\text{where } ab = cd, \quad a, b, c, d = 0, 1, 2, \ldots, \qquad (10.22c)$$

and in particular,

$$P_p^{(n)} \equiv P_{p^n} \pmod p, \quad n = 0, 1, 2, \ldots. \qquad (10.22d)$$

The Sierpinsky Gasket and Carpet

Figure 10.9 is obtained from matrix (10.8b) by replacing 1 by a little gray triangle, and 0 by a similar white triangle, then skewing the large triangle until it is symmetrical with respect to the vertical axis. The resulting form is referred to as the *Sierpinsky gasket*. The gasket can also be obtained, as shown in figure 10.10, by successively carving smaller and smaller triangles out of the initial solid triangle. Figure 10.11 is the geometric transpose of (P_{27} mod3).

Fig. 10.10. Geometric construction of the Sierpinsky gasket, following successive divisions of the triangle.

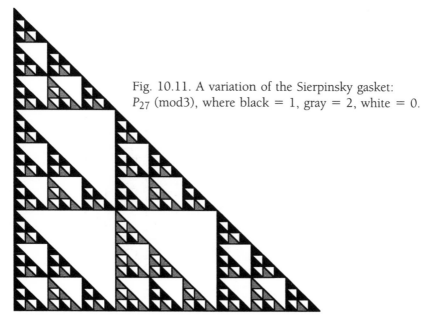

Fig. 10.11. A variation of the Sierpinsky gasket: P_{27} (mod3), where black = 1, gray = 2, white = 0.

Another interesting figure invented by Sierpinsky is shown in figure 10.12b. It is usually obtained by carving out squares from a large initial square. It can also be generated using the Kronecker product procedure, starting with matrix S of figure 10.12a. Infinite variations can be invented on the theme of the Sierpinsky carpet, one of which is shown in figures 10.13a and b. Figures 10.14a and b show yet another variation.

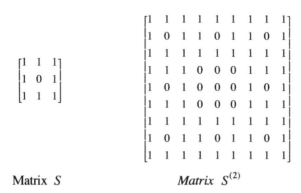

$$\begin{bmatrix} 1 & 1 & 1 \\ 1 & 0 & 1 \\ 1 & 1 & 1 \end{bmatrix}$$

$$\begin{bmatrix} 1 & 1 & 1 & 1 & 1 & 1 & 1 & 1 & 1 \\ 1 & 0 & 1 & 1 & 0 & 1 & 1 & 0 & 1 \\ 1 & 1 & 1 & 1 & 1 & 1 & 1 & 1 & 1 \\ 1 & 1 & 1 & 0 & 0 & 0 & 1 & 1 & 1 \\ 1 & 0 & 1 & 0 & 0 & 0 & 1 & 0 & 1 \\ 1 & 1 & 1 & 0 & 0 & 0 & 1 & 1 & 1 \\ 1 & 1 & 1 & 1 & 1 & 1 & 1 & 1 & 1 \\ 1 & 0 & 1 & 1 & 0 & 1 & 1 & 0 & 1 \\ 1 & 1 & 1 & 1 & 1 & 1 & 1 & 1 & 1 \end{bmatrix}$$

Matrix S *Matrix $S^{(2)}$*

Fig. 10.12a. Matrix for the Sierpisky carpet.

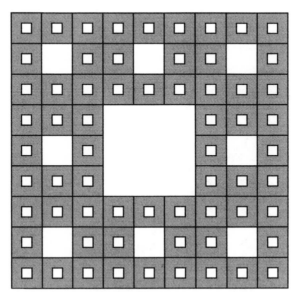

Fig. 10.12b. The Sierpinsky carpet.

$$\begin{bmatrix} 1 & 1 & 1 \\ 1 & 2 & 1 \\ 1 & 1 & 1 \end{bmatrix}$$

Matrix G

$$\begin{bmatrix} 1 & 1 & 1 & 1 & 1 & 1 & 1 & 1 & 1 \\ 1 & 2 & 1 & 1 & 2 & 1 & 1 & 2 & 1 \\ 1 & 1 & 1 & 1 & 1 & 1 & 1 & 1 & 1 \\ 1 & 1 & 1 & 2 & 2 & 2 & 1 & 1 & 1 \\ 1 & 2 & 1 & 2 & 1 & 2 & 1 & 2 & 1 \\ 1 & 1 & 1 & 2 & 2 & 2 & 1 & 1 & 1 \\ 1 & 1 & 1 & 1 & 1 & 1 & 1 & 1 & 1 \\ 1 & 2 & 1 & 1 & 2 & 1 & 1 & 2 & 1 \\ 1 & 1 & 1 & 1 & 1 & 1 & 1 & 1 & 1 \end{bmatrix}$$

Matrix $G^{(2)}$ (mod 3)

Fig. 10.13a. Matrix for the modified Sierpinsky carpet.

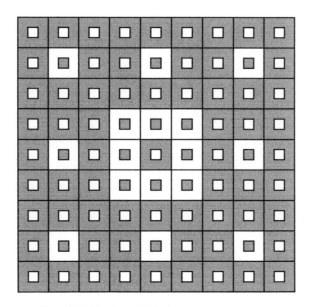

Fig. 10.13b. A modified Sierpinsky carpet.

$$
\begin{bmatrix}
1\,1\,1\,1\,1 & 1\,1\,1\,1\,1 & 1\,1\,1\,1\,1 & 1\,1\,1\,1\,1 & 1\,1\,1\,1\,1 \\
1\,2\,2\,2\,1 & 1\,2\,2\,2\,1 & 1\,2\,2\,2\,1 & 1\,2\,2\,2\,1 & 1\,2\,2\,2\,1 \\
1\,2\,0\,2\,1 & 1\,2\,0\,2\,1 & 1\,2\,0\,2\,1 & 1\,2\,0\,2\,1 & 1\,2\,0\,2\,1 \\
1\,2\,2\,2\,1 & 1\,2\,2\,2\,1 & 1\,2\,2\,2\,1 & 1\,2\,2\,2\,1 & 1\,2\,2\,2\,1 \\
1\,1\,1\,1\,1 & 1\,1\,1\,1\,1 & 1\,1\,1\,1\,1 & 1\,1\,1\,1\,1 & 1\,1\,1\,1\,1 \\
1\,1\,1\,1\,1 & 2\,2\,2\,2\,2 & 2\,2\,2\,2\,2 & 2\,2\,2\,2\,2 & 1\,1\,1\,1\,1 \\
1\,2\,2\,2\,1 & 2\,1\,1\,1\,2 & 2\,1\,1\,1\,2 & 2\,1\,1\,1\,2 & 1\,2\,2\,2\,1 \\
1\,2\,0\,2\,1 & 2\,1\,0\,1\,2 & 2\,1\,0\,1\,2 & 2\,1\,0\,1\,2 & 1\,2\,0\,2\,1 \\
1\,2\,2\,2\,1 & 2\,1\,1\,1\,2 & 2\,1\,1\,1\,2 & 2\,1\,1\,1\,2 & 1\,2\,2\,2\,1 \\
1\,1\,1\,1\,1 & 2\,2\,2\,2\,2 & 2\,2\,2\,2\,2 & 2\,2\,2\,2\,2 & 1\,1\,1\,1\,1 \\
1\,1\,1\,1\,1 & 2\,2\,2\,2\,2 & 0\,0\,0\,0\,0 & 2\,2\,2\,2\,2 & 1\,1\,1\,1\,1 \\
1\,2\,2\,2\,1 & 2\,1\,1\,1\,2 & 0\,0\,0\,0\,0 & 2\,1\,1\,1\,2 & 1\,2\,2\,2\,1 \\
1\,2\,0\,2\,1 & 2\,1\,0\,1\,2 & 0\,0\,0\,0\,0 & 2\,1\,0\,1\,2 & 1\,2\,0\,2\,1 \\
1\,2\,2\,2\,1 & 2\,1\,1\,1\,2 & 0\,0\,0\,0\,0 & 2\,1\,1\,1\,2 & 1\,2\,2\,2\,1 \\
1\,1\,1\,1\,1 & 2\,2\,2\,2\,2 & 0\,0\,0\,0\,0 & 2\,2\,2\,2\,2 & 1\,1\,1\,1\,1 \\
1\,1\,1\,1\,1 & 2\,2\,2\,2\,2 & 2\,2\,2\,2\,2 & 2\,2\,2\,2\,2 & 1\,1\,1\,1\,1 \\
1\,2\,2\,2\,1 & 2\,1\,1\,1\,2 & 2\,1\,1\,1\,2 & 2\,1\,1\,1\,2 & 1\,2\,2\,2\,1 \\
1\,2\,0\,2\,1 & 2\,1\,0\,1\,2 & 2\,1\,0\,1\,2 & 2\,1\,0\,1\,2 & 1\,2\,0\,2\,1 \\
1\,2\,2\,2\,1 & 2\,1\,1\,1\,2 & 2\,1\,1\,1\,2 & 2\,1\,1\,1\,2 & 1\,2\,2\,2\,1 \\
1\,1\,1\,1\,1 & 2\,2\,2\,2\,2 & 2\,2\,2\,2\,2 & 2\,2\,2\,2\,2 & 1\,1\,1\,1\,1 \\
1\,1\,1\,1\,1 & 1\,1\,1\,1\,1 & 1\,1\,1\,1\,1 & 1\,1\,1\,1\,1 & 1\,1\,1\,1\,1 \\
1\,2\,2\,2\,1 & 1\,2\,2\,2\,1 & 1\,2\,2\,2\,1 & 1\,2\,2\,2\,1 & 1\,2\,2\,2\,1 \\
1\,2\,0\,2\,1 & 1\,2\,0\,2\,1 & 1\,2\,0\,2\,1 & 1\,2\,0\,2\,1 & 1\,2\,0\,2\,1 \\
1\,2\,2\,2\,1 & 1\,2\,2\,2\,1 & 1\,2\,2\,2\,1 & 1\,2\,2\,2\,1 & 1\,2\,2\,2\,1 \\
1\,1\,1\,1\,1 & 1\,1\,1\,1\,1 & 1\,1\,1\,1\,1 & 1\,1\,1\,1\,1 & 1\,1\,1\,1\,1 \\
\end{bmatrix}
$$

Fig. 10.14a. Matrices H (top left corner) and $H^{(2)}$.

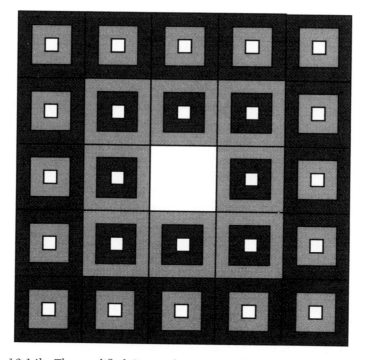

Fig. 10.14b. The modified Sierpinsky carpet, where dark gray = 1, light gray = 2, and white = 0.

The Cantor Dust

We must not admit the possibility of continuous observation.
Observations are to be regarded as discrete, disconnected
events. Between them there are gaps which we cannot fill in.
. . . Sometimes these events form chains that give the illusion
of permanent beings—but only in particular circumstances
and only for an extremely short period of time
in every single case.
(*Erwin Schrödinger*)[5]

[5] Quoted in Bell, *The World of Mathematics*, p. 1057.

Georg Cantor made an immense contribution to that conceptually difficult branch of mathematics, indeed, of human knowledge in the largest sense, known as infinity. A figure introduced by Cantor to illustrate the paradoxes of the infinite, known as the *Cantor dust*, is constructed as follows. Begin with the straight line shown in figure 10.15a at stage 0. That line is regarded as having unit length. It includes origin 0 and extends infinitely close to the 1 mark, without containing it. Remove its middle third part, which contains the 1/3 mark and extends infinitely close to the 2/3 mark, without containing it. We obtain the line at stage 1. Proceed with the following stages, every time removing the middle third of each remaining part together with its origin, but without its end point. At stage n, the gaps are $2^n - 1$ in number, and so are their untouched end points: 2/3 at stage 1, then 2/3, 2/9, 8/9 at stage 2, then 2/3, 2/9, 8/9, 2/27, 8/27, 20/27, 26/27 at stage 3, and so on. It therefore seems that as the line is gradually depleted from its points, more and more end points emerge.

As the process is extended to infinity, we are left with infinitely many end points, separated by infinitely many gaps, each containing infinitely many points. Stranger still, the remaining *dust* is of the same power as the continuum; it is not denumerable. To prove that point, we shall resort to the ternary, or triadic positional, number system. Figure 10.15b shows how a triadic yardstick can be constructed for the measurement of distances included between 0 and 1 in increments of $1/3^n$, where n is the number of stages. White corresponds to 0, gray to 1, and black to 2. To each stage corresponds a new digit position within the mantissa, to the right of the preceding positions. Stage 1 indicates that the coordinate of A is equal to or larger than 0, but smaller than 1/3, and the coordinate of B is equal to or larger than $(0.1)_3 = 1/3$, but smaller than 2/3. Stage 4 indicates that $(0.0121.)_3 \leq A < (0.0122)_3$ and $(0.1012)_3 \leq B < (0.1020)_3$. The reason for the $\leq$ and $<$ signs is that every interval contains its origin, but excludes its endpoint. Applying the yardstick to Cantor's figure, we discover that the "dust" does not contain any mantissa involving 1 within its digits. The remaining mantissas consist of zeros and twos in every possible configuration of infinite length. If we replace digit 2 by digit 1, and regard the resulting configurations as base-2 mantissas, we discover that every number between 0 and 1 is represented; in other words, the *continuum*. The remaining dust thus contains as many points as the original line!

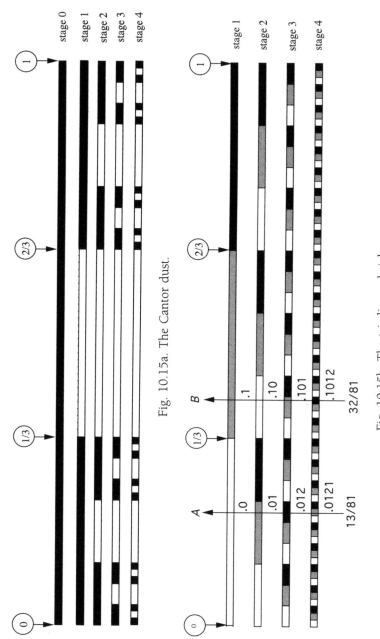

stage 0
stage 1
stage 2
stage 3
stage 4

Fig. 10.15a. The Cantor dust.

stage 1
stage 2
stage 3
stage 4

.1
.10
.101
.1012
32/81

.0
.01
.012
.0121
13/81

Fig. 10.15b. The triadic yardstick.

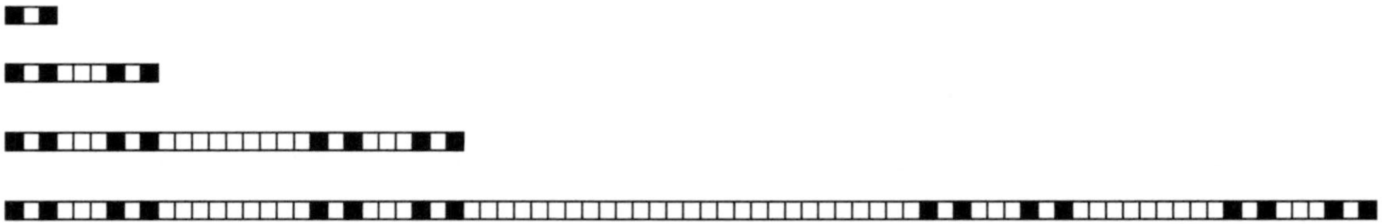

Fig. 10.15c. Cantor's dust is fractal.

γ :	0	1	2	3	4	5	6	7	8	9	10	11	12	13	14	15	16	17	18	19	20	21	22	23	24	25	26
$(\delta_0^\gamma)_3$:	0	1	2	0	1	2	0	1	2	0	1	2	0	1	2	0	1	2	0	1	2	0	1	2	0	1	2
$(\delta_1^\gamma)_3$:	0	0	0	1	1	1	2	2	2	0	0	0	1	1	1	2	2	2	0	0	0	1	1	1	2	2	2
$(\delta_2^\gamma)_3$:	0	0	0	0	0	0	0	0	0	1	1	1	1	1	1	1	1	1	2	2	2	2	2	2	2	2	2
		*			*			*			*			*			*			*			*			*	

Fig. 10.15d. Numbers γ not containing 1 in their base-3 representation.

Consider the following vector and its successive orders:

$$\vec{V} = [1\,0\,1],$$
$$\vec{V}^{(2)} = [1\,0\,1\,|\,0\,0\,0\,|\,1\,0\,1],$$
$$\vec{V}^{(3)} = [1\,0\,1\,0\,0\,0\,1\,0\,1\,|\,0\,0\,0\,0\,0\,0\,0\,0\,0\,|\,1\,0\,1\,0\,0\,0\,1\,0\,1].$$

Figure 10.15c shows the geometric transcription of the corresponding vectors. The process generates Cantor's dust, as it highlights its fractal character. That process is equivalent to that of figure 10.15a but proceeds in the opposite direction (adding new points to the line at every stage, instead of removing existing points). Figure 10.15d shows the process that consists of writing the numbers from 0 to $3^n - 1$, and removing those containing digit 1 in their representation.

The Thue-Morse Sequence and Tiling

In 1906 the Norwegian mathematician Axel Thue published a sequence that was later referred to as the Thue-Morse sequence. The sequence is obtained by replacing 1 by 0, and 2 by 1, in figure 16a, revealing its fractal character. The number whose base-2 mantissa is the infinite length Thue-Morse sequence is *transcendental*. Figure 10.16b shows how the sequence can be constructed geometrically. Every new sequence is obtained by adding to the previous sequence its "negative."

Following a similar process both horizontally and vertically, as in figure 10.16c, the tiling of figure 10.16d is obtained. Despite its symmetry with respect to the diagonals, and the apparent regularity of its motif, no stage can be found to regularly tile the following stage.

$\vec{T} =$ [1 2]

$\vec{T}^{(2)} (\mathrm{mod}\,3) =$ [1 2 * 2 1]

$\vec{T}^{(3)} (\mathrm{mod}\,3) =$ [1 2 2 1 * 2 1 1 2]

$\vec{T}^{(4)} (\mathrm{mod}\,3) =$ [1 2 2 1 2 1 1 2 * 2 1 1 2 1 2 2 1]

$\vec{T}^{(5)} (\mathrm{mod}\,3) =$ [1 2 2 1 2 1 1 2 2 1 1 2 1 2 2 1 * 2 1 1 2 1 2 2 1 1 2 2 1 2 1 1 2]

T.M Sequence = [0 1 1 0 1 0 0 1 1 0 0 1 0 1 1 0 * 1 0 0 1 0 1 1 0 0 1 1 0 1 0 0 1]

Fig. 10.16a. The Thue-Morse sequence is fractal.

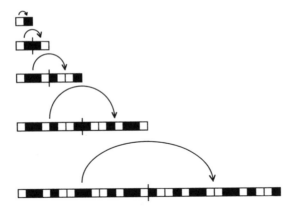

Fig. 10.16b. The first five orders of the Thue-Morse sequence and its geometric interpretatioin, where black = 1 and white = 0.

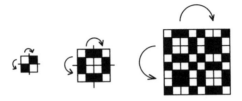

Fig. 10.16c. Generating successive tiling generations.

Fig. 10.16d. A fifth-order tiling inspired by the Thue-Morse sequence.

HIGHER-DIMENSIONAL LATTICES

So far, we have examined one- and two-dimensional lattices, namely, vectors and matrices. We have seen that the Kronecker product can be effected between a matrix and a vector, in other words, that the lattices at hand need not possess the same number of dimensions, when considered individually. When multiplied by a matrix in two-dimensional space, a vector must be regarded as a unidimensional matrix, consisting of either a single row or a single column. The same can be said of any d-dimensional lattice, when embedded in D-dimensional space, where $D \geq d$.

With that in mind, we can generalize the Kronecker product to any number of dimensions. We shall begin with figure 10.17a, in which the bottom-left diagram shows the three-dimensional system of coordinate axes. Indices μ, μ', μ'' of matrix M, and indices ρ, ρ', ρ'' of matrix R respectively lie upon axes, d, d', d''. Figure 10.17b shows Kronecker product $M \otimes R$, whose element $G_{\gamma,\gamma',\gamma'}$ is given by

$$G_{\gamma,\gamma',\gamma'} = (M \otimes R)_{(\mu+m\rho),(\mu'+m'\rho'),(\mu''+m''\rho')} = M_{\mu,\mu',\mu''} \times R_{\rho,\rho',\rho''}. \quad (10.23)$$

As the number of lattices or the number of dimensions grows, it becomes increasingly difficult to calculate the value of the Kronecker product elements. The following tabular method, illustrated by means of an example involving three four-dimensional lattices, has the merit of organizing the process, and avoiding errors. Indices γ, γ', γ'', γ''' are written as shown, along with their representations within their respective bases:

G	S	R	M
γ base$(s, r, m.) =$	σ	ρ	μ
γ' base$(s', r', m'.) =$	σ'	ρ'	μ'
γ'' base$(s'', r'', m''.) =$	σ''	ρ''	μ''
γ''' base$(s''', r''', m'''.) =$	σ'''	ρ'''	μ'''

The indices of M, R, S are then found beneath their respective symbols, in ascending order.

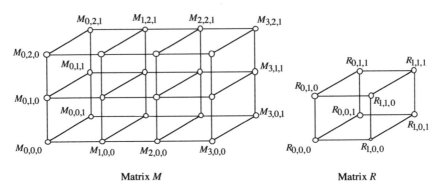

Matrix *M*   Matrix *R*

Fig. 10.17a. Three-dimensional matrices M and R.

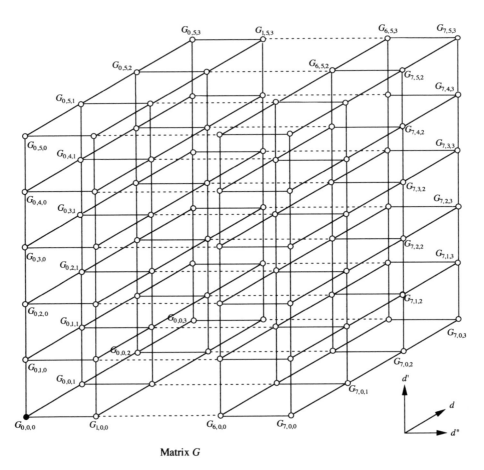

Matrix *G*

Fig. 10.17b. Three-dimensional Kronecker product $G = M \otimes R$.

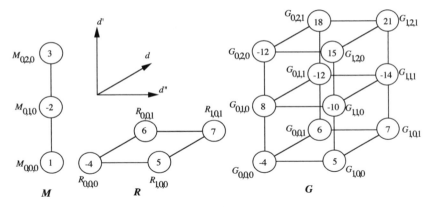

Fig. 10.18. Kronecker product of a matrix and a perpendicular vector.

Commutativity and Higher Dimensions

Figure 10.18 illustrates the commutativity of the Kronecker product of a vector and a matrix, in three-dimensional space, where

$$m = m'' = 1, \qquad r' = 1, \quad \text{that is,} \quad \mu = \mu'' = 0, \rho' = 0,$$

hence

$$G_{\mu+m\rho,\mu'+m'\rho',\mu''+m''\rho'} = G_{\rho,\mu',\rho'} = (M \otimes R)_{\rho,\mu',\rho'} = (R \otimes M)_{\rho,\mu',\rho'} = M_{0,\mu',0}R_{\rho,0,\rho'}.$$

Generally, in any pluridimensional space, the Kronecker product of two lattices is commutative if and only if they share no common dimension. In three-dimensional space, that happens with two vectors or with a vector and a matrix. In four-dimensional space, the product of two matrices can be commutative as well. In the case of more than two lattices, no lattice pair can share a dimension. These statements obviously do not account for the commutativity of the product of two different orders of the same lattice.

The Three-Dimensional Sierpinsky Pyramid and Menger Sponge

Consistent with formulation (10.23), figure 10.19a and plate 24 show the construction of the three-dimensional Sierpinsky gasket, in the shape of a regular tetrahedron riddled with holes. Figure 10.19a shows the structure of the pyramid's seed, where seven out of the twenty-seven vertices of lattice A are chosen to contain 1, while the remaining vertices contain 0. Thus,

$$A_{0,0,0} = A_{1,0,0} = A_{1,0,1} = A_{2,0,0} = A_{2,0,2} = A_{1,1,0} = A_{2,2,0} = 1.$$

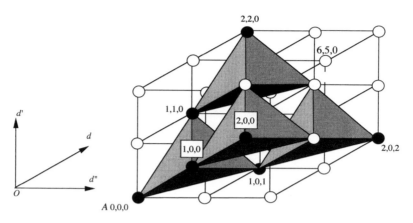

Fig. 10.19a. The pyramid's seed, consisting of three
tetrahedral pyramidions.

Plate 24 shows the construction of $A^{(3)}$. With each new order, every
pyramid spews out an identical pyramid from each of its free black
vertices, corresponding to $A_{\gamma,\gamma',\gamma''} = 1$.

Examples

	δ_2	δ_1	δ_0		δ_2	δ_1	δ_0		δ_2	δ_1	δ_0
$\gamma = 4$	1	0	0	$\gamma = 7$	1	1	1	$\gamma = 7$	1	1	1
$\gamma' = 3$	0	1	1	$\gamma' = 3$	0	1	1	$\gamma' = 4$	1	0	0
$\gamma'' = 1$	0	0	1	$\gamma'' = 4$	1	0	0	$\gamma'' = 3$	0	1	1

$$A_{4,3,1} = A_{0,1,1} \times A_{0,1,0} \times A_{1,0,0} = 0 \times 0 \times 1 = 0,$$
$$A_{7,3,4} = A_{1,1,0} \times A_{1,1,0} \times A_{1,0,1} = 1 \times 1 \times 1 = 1,$$
$$A_{7,4,3} = A_{1,0,1} \times A_{1,0,1} \times A_{1,1,0} = 1 \times 1 \times 1 = 1.$$

An interesting exercise consists in attempting to visualize the shape of
the hole inside the fractal pyramid. The inner space between the seed's
pyramidions is an octagon. Figure 10.19b shows an isometric view of
the exploded pyramid, together with the inscribed virtual octagon. The
octagon's faces are shown one at a time in figure 10.19c. Faces 1, 3,
5, 7, which we might refer to as *solid*, also belong to the pyramidions,
while faces 2, 4, 6, 8, are referred to as *virtual*—are in the air, so to
speak.

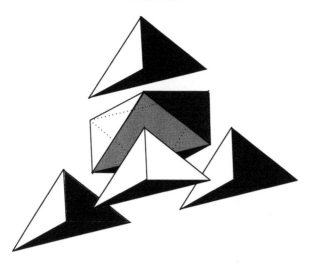

Fig. 10.19b. The exploded pyramid, showing its octagonal inner space.

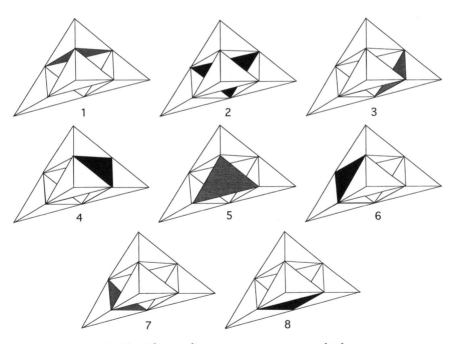

Fig. 10.19c. The seed's inner space is an octahedron.

229

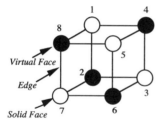

Fig. 10.19d. Interconnection of the octagon's faces in a first order pyramid.

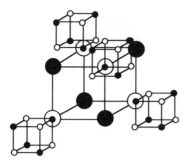

Fig. 10.19e. Interconnection of the octagons' faces in a second-order pyramid.

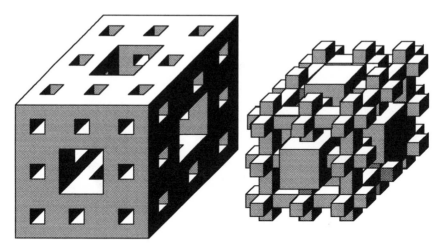

Fig. 10.19f. The three-dimensional Menger sponge and its fractal inside.

Figure 10.19d shows the octagon's faces, each represented by a small bubble. A line connecting two adjacent bubbles indicates that the corresponding faces of the octagon share a common edge. That exercise corroborates the observation that the center points of a regular octahedron coincide with the vertices of a cube (and conversely, the center points of a cube's faces coincide with the vertices of a regular octagon). The reader may attempt to visualize the inner space's shape for a second-order pyramid. Dealing with cubes is easier than with octagons, hence the diagram of figure 10.19e, in which a little solid bubble within a larger clear bubble corresponds to the sprouting of a little octagon's virtual face from its predecessor's solid face. The reader will discover that the pyramid's inner space is also fractal.

Figure 10.19f, which represents a third-order Menger sponge, is obtained from the Sierpinsky carpet by adding the third dimension, in the same manner as in the previous exercise. The figure also shows the inner space of a second-order sponge.

THE KRONECKER PRODUCT WITH RESPECT TO OTHER OPERATIONS

So far, we have examined the Kronecker product only with respect to arithmetic multiplication. Nothing prevents us, however, from defining product lattices with respect to other operations. For example, if the operation at hand is arithmetic addition, we have

$$\vec{V} = (1\ 2\ -3),$$
$$\vec{V}^{(2)} = (2\ 3\ -2\,|\,3\ 4\ -1\,|\,-2\ -1\ -6) \quad \text{(w.r.t. addition)}.$$

An interesting example is the following:

$(0\ 1) \otimes (0\ 2) = (0\ 1\ 2\ 3)$ (w.r.t. addition),

$(0\ 1) \otimes (0\ 2) \otimes (0\ 4) = (0\ 1\ 2\ 3\ 4\ 5\ 6\ 7),$

$(0\ 1) \otimes (0\ 2) \otimes (0\ 4) \otimes (0\ 8) = (0\ 1\ 2\ 3\ 4\ 5\ 6\ 7\ 8\ 9\ 10\ 11\ 12\ 13\ 14\ 15), \ldots .$

The addition's *neutral element* is 0, meaning that for any number a, we have $a + 0 = 0 + a = a$. A lattice A, whose *origin* $A_{0,0,0,\ldots}$ is the neutral element with respect to some operation, will generate *fractal lattice* $A^{(n)}$, $n > 1$, *with respect to that operation*, as matrix A will always find itself embedded, unchanged, near the lattice's origin, as in the previous example.

Another example is provided by the *Thue-Morse* sequence:

$$\vec{T} = (0\ 1),$$
$$\vec{T}^{(2)} = (0\ 1\ 1\ 0),$$
$$\vec{T}^{(3)} = (0\ 1\ 1\ 0\ 1\ 0\ 0\ 1) \quad \text{(w.r.t. modulo 2 addition)}.$$

A further example is the following:

$$\vec{S} = (0\ 1\ 2\ 3),$$
$$\vec{S}^{(2)} = (0\ 1\ 2\ 0\ 1\ 2\ 0\ 1\ 2\ 0\ 1\ 2\ 0\ 1\ 2\ 0),$$
$$\vec{S}^{(n)} = (0\ 1\ 2\ 0\ 1\ 2\ 0\ 1\ 2\ldots) \quad \text{(w.r.t. modulo 3 addition)}.$$

In general, for

$$\vec{S} = (0\ 1\ 2\ 3\ldots m-1), \text{ we get}$$
$$\vec{S}^{(n)} = (0\ 1\ 2\ldots(m-2)\ 0\ 1\ 2\ (m-2)\ 0\ 1\ 2\ldots)$$
$$\text{(w.r.t. modulo } (m-1) \text{ addition)}.$$

In other words, $S_\gamma^{(n)} \equiv \gamma\ (\mathrm{mod}(m-1))$ for any value of n. We can explain this as follows:

$$\vec{S}_\gamma^{(n)} \equiv \sum_{i=0}^{n-1} \vec{S}_{(\delta_i^\gamma)_m}\ (\mathrm{mod}(m-1)),$$

and since $\vec{S}_{(\delta_i^\gamma)_m} = (\delta_i^\gamma)_m$, we get $\vec{S}_\gamma^{(n)} \equiv \sum_{i=0}^{n-1} (\delta_i^\gamma)_m (\mathrm{mod}(m-1))$.

In chapter IX we learned that the modulo $(m-1)$ sum of the base-m digits of an integer is equal to that integer's modulo $(m-1)$ residue. Therefore, $S_\gamma^{(n)} \equiv \gamma\ (\mathrm{mod}(m-1))$ for any value of n.

Other operations can be imagined, such as that defined by the following table, which reflects its commutative character, as well as the existence of a neutral element, namely, 1. Indeed, for any element a, we have $a \oplus 1 = 1 \oplus a = a$:

⊕	0	1	2
0	1	0	2
1	0	1	2
2	2	2	0,

and we get, for example,

$$\left.\begin{array}{l} \vec{V} = (1\ 2\ 0), \\ \vec{V}^{(2)} = (1\ 2\ 0\ 2\ 0\ 2\ 0\ 2\ 1) \end{array}\right\} \quad \text{(w.r.t. the operation } \oplus\text{)}.$$

Obviously, the operation under consideration need not be commutative. Remember that the Kronecker product itself is not commutative.

FRACTAL LINKAGES

Imagine an old-fashioned surveyor chain, made up of m articulated straight segments of unit length. When the chain is fully stretched, its links can be likened to little arrows, all pointing in the same direction. Imagine also, drawn on the ground, a straight line segment, subdivided into l subsegments, each of which is equal to the chain's link. We acquire two solid pegs and drive them into the ground at the line's extremities, as in figure 10.20a, where $l = 3$, $m = 4$. The straight span between the pegs is called the *initiator*, and the chain segment is called the *generator* or *seed*. A geometric form such as that of the generator, all of whose links are of equal length, will be called a *regular linkage*. Otherwise, it will be referred to as an *irregular linkage*. The number m of links in the generator is called the *base*.

Turning to figure 10.20b, we define the *elementary vectors* represented by little arrows for $d = 3, 4, 5, 6, \ldots$ where j is the imaginary quantity $\sqrt{-1}$. To each value of d, which is referred to as the *modulus*, correspond d vectors e^0, e^ϕ, $e^{2\phi}$, $\ldots$ $e^{(d-1)\phi}$. Clearly, $e^{d\phi} = e^0 = 1$, $e^{(d+1)\phi} = e^\phi$, $\ldots$, and in general $e^{q\phi} = e^{(q \bmod d)\phi}$.

generator
initiator

Fig. 10.20a. The initiator and generator.

233

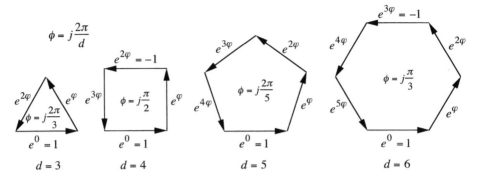

Fig. 10.20b. Elementary vectors.

The Koch Curve

So, Nat'ralists observe, a Flea Hath smaller Fleas that
on him prey. And these have smaller Fleas to bite 'em,
And so proceed ad infinitum.
(Jonathan Swift)

The generator of Figure 10.20a can now be represented by the vector

$$\vec{V} = (e^0, e^\phi, e^{5\phi}, e^0), \quad \text{where } d = 6, m = 4, \phi = j\frac{\pi}{3}, \qquad (10.24a)$$

or by

$$(e^0, -e^{2\phi}, -e^\phi, e^0), \quad \text{where } d = 3, m = 4, \phi = j\frac{2\pi}{3}. \qquad (10.24b)$$

If we consider only the factors of ϕ in sequence (10.24a), we can write

$$\vec{\Phi} = (0\ 1\ 5\ 0),$$
$$\vec{\Phi}^{(2)} = (0\ 1\ 5\ 0\ 1\ 2\ 0\ 1\ 5\ 0\ 4\ 5\ 0\ 1\ 5\ 0) \quad \text{(w.r.t. mod 6 addition)},$$
$$\cdots \qquad\qquad\qquad\qquad\qquad\qquad (10.25)$$

with the understanding that any element $K_\gamma^{(n)}$ indicates that the corresponding link forms an angle $\vec{\Phi}_\gamma^{(n)}$ ($\pi/3$) radians with respect to the horizontal axis, where angles are measured anticlockwise. That angle will be referred to as the link's *argument*, and vector $\vec{\Phi}$ as the *argument generator*. In the general case,

$$\vec{\Phi}_\gamma^{(n)} = \sum_{i=0}^{n-1} \vec{\Phi}_{(\delta_i^\gamma)_m} \ (\text{mod } d), \tag{10.26}$$

where $(\delta_i^\gamma)_m$ is the ith digit, base m, of γ, and $0 \le \gamma \le m^n - 1$.

In the case of generator (10.25), the argument of link number γ is equal to

$$\frac{\pi}{3} \vec{\Phi}_\gamma^{(n)} = \frac{\pi}{3} \left(\sum_{i=0}^{n-1} \vec{\Phi}_{(\delta_i^\gamma)_4} \ (\text{mod } 6) \right). \tag{10.27}$$

For example, the inclination of link 55, shown in figure 10.21, is

$$\frac{\pi}{3} \vec{\Phi}_{55}^{(n)} = \frac{\pi}{3} \left(\sum_{i=0}^{2} \vec{\Phi}_{(\delta_i^{55})_4} \ (\text{mod } 6) \right).$$

And since $55 = (3\ 1\ 3.)_4$, we get

$$\frac{\pi}{3} \vec{\Phi}_{55}^{(3)} = \frac{\pi}{3} (\vec{\Phi}_3 + \vec{\Phi}_1 + \vec{\Phi}_3) = \frac{\pi}{3} (0 + 1 + 0) = \frac{\pi}{3}.$$

(Observe that $\vec{\Phi}_\gamma^{(n')} = \vec{\Phi}_\gamma^{(n)}$ for any value of $n' \ge n$, $n = 0, 1, 2, \ldots, 0 \le \gamma < m^n - 1$).

The linkage whose first three stages are shown in figure 10.21a is known as the *Koch curve*. The *Koch snowflake* in figure 10.21b is obtained by replacing each side of an equilateral triangle by the Koch seed, then endlessly replacing each straight segment of the resulting figure by that seed. With every new iteration, the Koch curve's perimeter is multiplied by 4/3. An infinite number of iterations thus results in a perimeter of infinite length and leads to the paradoxical situation where a finite area is made to fit within a closed curve of infinite perimeter! That paradox has baffled mathematicians, leading them to refer to the curve as "monstrous" or "pathological," and Benoît Mandelbrot to refer to it as the

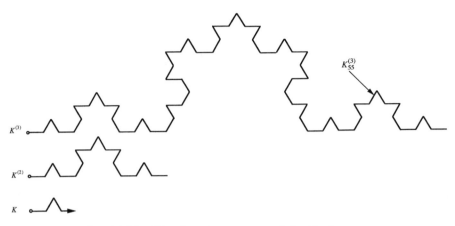

$K^{(3)}$

$K^{(2)}$

K

$K^{(3)}_{55}$

Fig. 10.21a. The first three stages of the Koch curve.

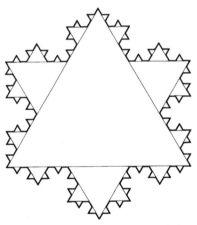

Fig. 10.21b. The Koch snowflake.

Chimeric Island. (In Greek mythology, a chimera was a fire-breathing female monster with a lion's head, a dragon's tail, and a goat's belly.) It belongs to those curves that, following an infinite number of stages, have no tangent at any point. It will be observed that the pattern of the three stages is the same, namely, the seed itself, and the building block of any stage is the entire previous stage.

236

The Peano Space-Filling Curve

Is it possible to establish a one-to-one correspondence between the points of a square and those of a line? In other words, is the power of the square's points that of the continuum? That question baffled mathematicians until Giuseppe Peano (1858–1932)[6] and David Hilbert (1862–1943)[6] discussed a strange wiggly curve that achieves the unexpected performance of exploring every point inside a given square (Peano in 1890 and Hilbert in 1891). That regular linkage, whose argument generator is shown in figure 10.22a and its second-order in figure 10.22b, is fractal (corners are rounded for clarity). The curve in figure 10.22a is inscribed within a square that is used as an oriented building block for the construction of the larger square in Figure 10.22b. That square, in turn, is used to construct a large square, and so on.

$$d = 4, \; m = 9, \; \overrightarrow{\Phi} = (0\;1\;0\;3\;2\;3\;0\;1\;0)$$

Fig. 10.22a. The Peano curve generator.

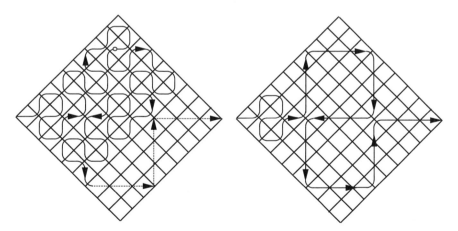

Fig. 10.22b. Exploring the square with the Peano curve (the square of fig. 10.22a is enlarged three times).

[6] "Sur une courbe qui remplit toute une aire plane," *Mathematische Annalen* 36 (1890), pp. 157–60.

Upon constructing the larger square, the building blocks are oriented following the same pattern as that of the generator. Let us posit that the square's side measures $1/3\sqrt{2}$ units in figure 10.22a. That makes the length of the curve equal to 1. We arbitrarily place upon the curve a little round dot at a distance of, say, 8/27 from the origin, measured along the curve.

Turning to figure 10.22b, a little dot is also placed at a distance of 8/27 from the origin. If the figure's size is reduced to that of figure 10.22a, the two dots will almost coincide. Continuing the process, every new dot will very nearly coincide with its predecessor. Proceeding to infinity, the successive dots converge to an "attractor," singularly corresponding to the fraction 8/27. To every fractional part of 1 thus corresponds one and only one point within the square, establishing the sought-after one-to-one correspondence.

A Collection of Regular Fractal Linkages

Figures 10.23 through 10.30 show a sampling of fractal linkages that are essentially variations on the Koch curve.

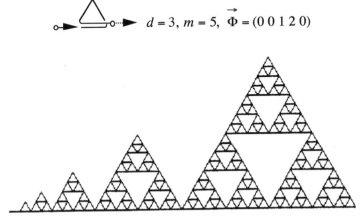

$d = 3,\ m = 5,\ \vec{\Phi} = (0\ 0\ 1\ 2\ 0)$

Fig. 10.23. Giza pyramids. A single Sierpinsky triangle whose size grows with every order can be obtained with $d = 3$, $m = 5$, $\vec{\Phi}_1, = (0\ 1\ 2\ 0\ 0)$.

238

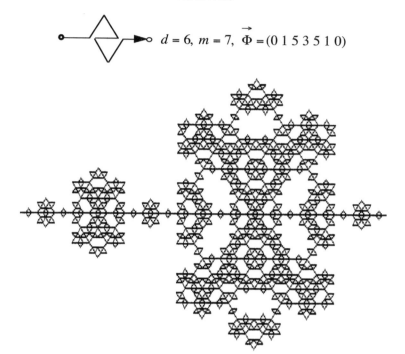

$d = 6, \; m = 7, \; \vec{\Phi} = (0\;1\;5\;3\;5\;1\;0)$

Fig. 10.24. Snow crystal.

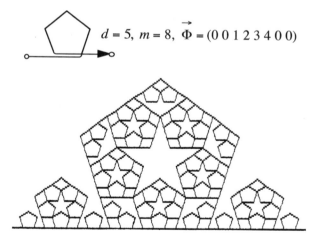

$d = 5, \; m = 8, \; \vec{\Phi} = (0\;0\;1\;2\;3\;4\;0\;0)$

Fig. 10.25. Pentagonal variation no. 1 on the Koch curve.

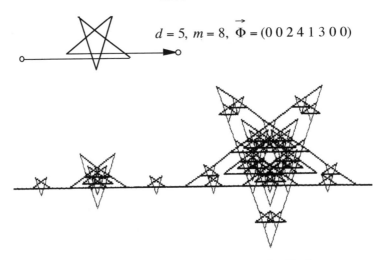

$d = 5$, $m = 8$, $\vec{\Phi} = (0\ 0\ 2\ 4\ 1\ 3\ 0\ 0)$

Fig. 10.26. Pentagonal variation no. 2 on the Koch curve.

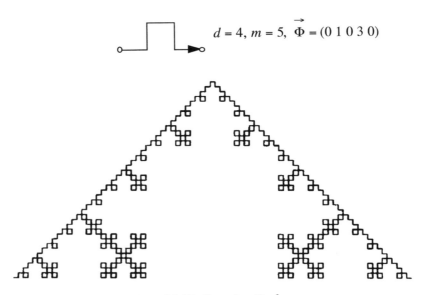

$d = 4$, $m = 5$, $\vec{\Phi} = (0\ 1\ 0\ 3\ 0)$

Fig. 10.27. Squaring Koch.

$d = 6, \ m = 10, \ \vec{\Phi} = (0\ 1\ 2\ 1\ 1\ 5\ 5\ 4\ 5\ 0)$

Fig. 10.28. Christmas tree.

$d = 4, \; m = 2, \; \vec{\Phi} = (0\ 1)$

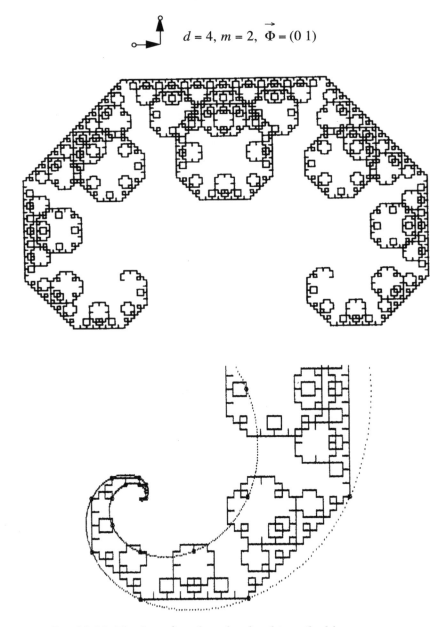

Fig. 10.29. The Levy fractal, and a detail inscribed between two logarithmic spirals.

$d = 6, m = 5, \overrightarrow{\Phi} = (0\ 1\ 2\ 3\ 4)$

Fig. 10.30. Cabbage.

Mixed Regular Linkages and Corresponding Tesselations

Whereas the linkages described so far were obtained by higher and higher orders of a single generating vector, the next few linkages (figs. 10.31–10.34) are obtained using two different vectors, the first of which, namely, $\vec{V}_1$, is referred to as the seed, and the other, namely, $\vec{V}_2$, as the template. Both vectors share the same modulus d, and the linkage is constructed according to $\vec{V}_1 \otimes \vec{V}_2^{(n)}$. If the corresponding argument generators are $\vec{\Phi}_1$, $\vec{\Phi}_2$, the linkage is constructed according to

$$\vec{\Phi}_1 \otimes \vec{\Phi}_2^{(n)} \quad \text{(w.r.t. mod } d \text{ addition)}.$$

$$d = 6, \quad \vec{\Phi}_1 = (0\ 1\ 0\ 4\ 4\ 0\ 1\ 0), \quad \vec{\Phi}_2 = (0\ 1\ 2\ 3\ 4\ 5\ 6\ 7\ 8\ 9\ 10)$$

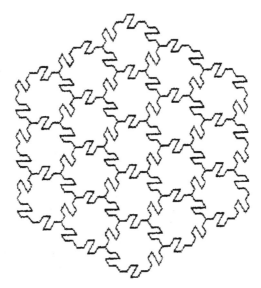

Fig. 10.31. Gears.

$$d = 6, \quad \vec{\Phi}_1 = (0\ 1\ 5\ 5\ 1\ 0), \quad \vec{\Phi}_2 = (0\ 1\ 2\ 3\ 4\ 5\ 6\ 7)$$

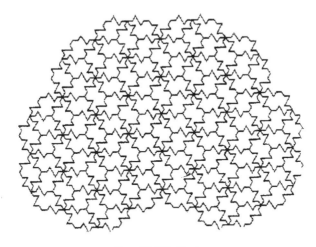

Fig. 10.32. Brain.

$$d = 6, \quad \vec{\Phi}_1 = (0\ 2\ 2\ 0\ 5\ 4\ 0)), \quad \vec{\Phi}_2 = (0\ 1\ 2\ 3\ 4\ 5\ 6\ 7\ 8\ 9)$$

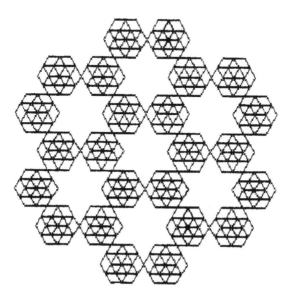

Fig. 10.33. Star of David.

$$d = 8, \quad \vec{\Phi_1} = (0\ 1\ 0\ 3\ 2\ 3\ 0\ 1\ 0), \quad \vec{\Phi_2} = (0\ 1\ 2\ 3\ 4\ 5\ 6\ 7)$$

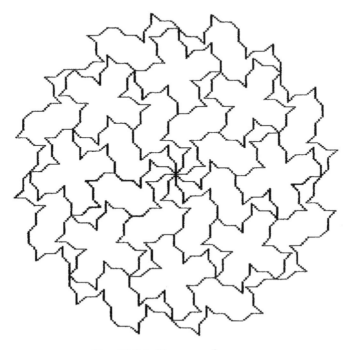

Fig. 10.34. Horses and crosses.

An Irregular Fractal Linkage: The Pentagonal "Eiffel Tower"

The linkages examined so far were regular, meaning that their elementary links were of equal length. An irregular fractal linkage that bears a very close resemblence to Pythagoras's lute,[7] and that we refer to as the pentagonal Eiffel Tower, is obtained using the generating vector above the figure. The tower is constructed horizontally then vertically erected (fig. 10.35).

---

[7] I first discovered the name "Pythagoras's lute" in Jan Gullberg's delightful *Mathematics: From the Birth of Numbers* (New York: W.W. Norton, 1996), p. 420.

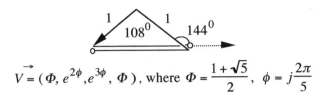

$$\vec{V} = (\Phi, e^{2\phi}, e^{3\phi}, \Phi), \text{ where } \Phi = \frac{1+\sqrt{5}}{2}, \quad \phi = j\frac{2\pi}{5}$$

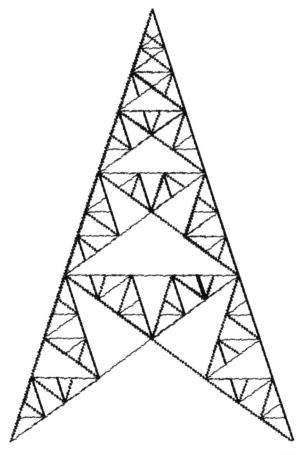

Fig. 10.35. An irregular fractal linkage: The pentagonal "Eiffel Tower."

Appendix: Simplifying Symbols

When the number of dimensions is large, it becomes unwieldy to write γ, γ', γ'', γ^{IV}, ... The following unambiguous symbols can therefore be defined, which greatly simplify notation.

We shall write

$$\{\mu\} = \mu, \mu', \mu'', \mu''', \ldots, \{\rho\} = \rho, \rho', \rho'', \rho''', \ldots,$$

$$\{\sigma\} = \sigma, \sigma', \sigma'', \sigma''', \ldots, \{\gamma\} = \gamma, \gamma', \gamma'', \gamma''', \ldots.$$

The equation

$$G_{\gamma,\gamma',\gamma''} = (M \otimes R \otimes S)_{(\mu+m\rho+mr\sigma),(\mu'+m'\rho'+m'r'\sigma'),(\mu''+m''\rho''+m''r''\sigma'')}$$

$$= M_{\mu,\mu',\mu''} \, R_{\rho,\rho',\rho''} \, S_{\sigma,\sigma',\sigma''}$$

can therefore be rewritten as

$$G_{\{\gamma\}} = (M \otimes R \otimes S)_{\{\mu+m\rho+mr\sigma\}} = M_{\{\mu\}} \, R_{\{\rho\}} \, S_{\{\sigma\}}.$$

We can also write

$$\{\delta_0^\gamma\} \equiv \delta_0^\gamma, \delta_0^{\gamma'}, \delta_0^{\gamma''}, \delta_0^{\gamma'''}, \ldots, \{\delta_1^\gamma\} \equiv \delta_1^\gamma, \delta_1^{\gamma'}, \delta_1^{\gamma''}, \delta_1^{\gamma'''}, \ldots,$$

and in general

$$\{\delta_0^\gamma\} \equiv \delta_i^\gamma, \delta_i^{\gamma'}, \delta_i^{\gamma''}, \delta_i^{\gamma'''}, \ldots,$$

where it is understood that

$$\delta_i^\gamma = (\delta_i^\gamma)_{(\ldots,m_2,m_1,m_0)} \quad \delta_i^{\gamma'} = (\delta_i^{\gamma'})_{(\ldots,m'_2,m'_1,m'_0)} \quad \delta_i^{\gamma''} = (\delta_i^{\gamma''})_{(\ldots,m''_2,m''_1,m''_0)} \cdots.$$

This allows us to write $M_{(0)\delta_0^\gamma,\delta_0^{\gamma'},\delta_0^{\gamma''},\ldots} = M_{(0)\{\delta_0^\gamma\}}$, $M_{(1)\delta_1^\gamma,\delta_1^{\gamma'},\delta_1^{\gamma''},\ldots} = M_{(1)\{\delta_1^\gamma\}}$, and in general $M_{(i)\delta_i^\gamma,\delta_i^{\gamma'},\delta_i^{\gamma''},\ldots} = M_{(i)\{\delta_i^\gamma\}}$, which can be further simplified as $M_{\{\delta_i^\gamma\}}$, by virtue of the redundant character of index (i). We can therefore replace the statement

$$G_{\gamma,\gamma',\gamma'',\ldots} = \prod_{i=0}^{n-1} M_{(i)\delta_i^\gamma,\delta_i^{\gamma'},\delta_i^{\gamma''},\ldots}$$

by the simpler statement

$$G_{\{\gamma\}} = \prod_{i=0}^{n-1} M_{\{\delta_i^{\gamma}\}},$$

which is identical to the statement defining the Kronecker product of n vectors, except for the brackets surrounding the indices.

That simplified notation may be useful when studying *tensors*. But that is another story.

Sweet is the lore which Nature brings;
Our meddling intellect
Mis-shapes the beauteous form of things;
We murder to dissect.

Enough of Science and Art;
Close up the barren leaves
Come forth and bring with you a heart
That watches and receives.

(William Wordsworth)

Index